User-Centered Requirements:
The Scenario-Based Engineering Process

User-Centered Requirements: The Scenario-Based Engineering Process

Karen L. McGraw
Cognitive Technologies, Annapolis, Maryland

Karan Harbison
University of Texas at Arlington

CRC Press
Taylor & Francis Group
Boca Raton London New York

CRC Press is an imprint of the
Taylor & Francis Group, an **informa** business

First Published by
Lawrence Erlbaum Associates, Inc., Publishers
10 Industrial Avenue
Mahwah, New Jersey 07430

Transferred to Digital Printing 2009 by CRC Press
6000 Broken Sound Parkway, NW Suite 300, Boca Raton, FL 33487
270 Madison Avenue, New York, NY 10016
2 Park Square, Milton Park, Abingdon, Oxon, OX14 4RN, UK

Cover design by Kathryn Houghtaling

Library of Congress Cataloging-in-Publication Data

McGraw, Karen L.
User-centered requirements : the scenario-based engineering process
/ Karen L. McGraw and Karan Harbison.
      p. cm.
    Includes bibliographical references and index.
    ISBN 0-8058-2064-7 (c : alk. paper). — ISBN 0-8058-
2065-5 (p : alk. paper)
    1. Systems engineering. 2. Systems analysis. I. Harbison,
Karan. II. Title.
   TA168.M35   1997
   005.1'2—dc20   96–31004
                                   CIP

Publisher's Note
The publisher has gone to great lengths to ensure the quality of this reprint
but points out that some imperfections in the original may be apparent.

Contents

Acknowledgments

Motivating yourself to write a book is tough. Although you know your processes, techniques, and experiences would be useful to others, it usually takes an external motivation to stimulate you to write about it. Our motivation for this text came not just from seeing the process and techniques work, resulting in better products that were accepted by the user community, but also from our colleagues, clients, and students. We thank them for continually asking and prodding.

Writing a book is difficult if your goal is to present and explain a process and techniques that you use daily. You have witnessed, firsthand, the difference the process and approach makes in the quality and completeness of your engineering products. You have seen differences in the enthusiasm of the intended users as they interact with you during the project and respond to the end result. You have used and taught hundreds of others to select and use the knowledge acquisition, elicitation, and analysis techniques. Consequently, your beliefs about processes and techniques, and the way you employ them, are difficult to explain in writing. To ensure that the techniques in this book are usable and that they can be replicated by the analyst/engineer reader, we asked selected colleagues to act as reviewers. The comments, changes, and additions they suggested made a difference.

We'd like to thank the following individuals, each of whom either stimulated the development of this book in some way, or actively reviewed it during development:

Col. John Silva, MD, Advanced Research Project Agency (SISTO)
Catherine Witt, RWD Technologies, Inc.
Dr. Jill Loukides, Cognitive Technologies
Lisa Mantock, ScenPro
Janie Nugent, Loral WDL
Kate Blodgett, RWD Technologies, Inc.
Bruce McGraw, Bell Atlantic
Sylita Smith, Federal Express
Dave Barnwell, Federal Express
John Alden, TerraQuest Metrics

Part I

INTRODUCTION
AND FOUNDATION

An Introduction to the Scenario-based Engineering Process

Computers and software programs are prevalent in our daily lives—and no longer simply as number crunching machines. Today they operate as decision aides that support complex human tasks. Furthermore, some advances in computer automation are enabling businesses to revise, enhance, and streamline long-standing business processes and roles. However, advances in the field of computer automation and consequent new uses for computer systems have outpaced developers' abilities to build systems that are responsive to users' needs.

Both systems/software engineering and business process re-engineering are providing partial responses to this problem. Systems/software engineering changes address the systems development aspect, whereas business process re-engineering activities address issues related to work process and role refinement, including cultural, organizational, and enterprise issues.

In this chapter we describe a process that marries these engineering practices, resulting in a synergy that addresses current needs in the areas of business process refinement, automation, and systems development. The chapter begins with a description of the systems development problem, followed by a brief chronology of existing systems/software engineering solutions. Next, we present a definition of the Scenario-based Engineering Process (SEP) and a discussion of the characteristics and guiding principles that define it. We then offer a more in-depth discussion of SEP as a model-based development process. Finally, we discuss the primary benefits of the SEP approach and present a rationale for its use.

This chapter is intended to provide the reader with an understanding of how the Scenario-based Engineering Process compares to existing solutions, and the contexts in which SEP is useful. It is designed to help readers meet the following goals:

- Recognize systems development problems that are not being solved with existing system/software engineering solutions.
- Build a framework for systems/software engineering against which to compare and understand SEP.
- Understand characteristics and guiding principles that define SEP.
- Identify the primary benefits of SEP.

THE SYSTEMS DEVELOPMENT PROBLEM

More than $250 billion is spent each year on information technology application development projects, with average project costs ranging from $400,000 for small companies, to over $2 million for large companies. As reported by the Standish Group (Johnson, 1995), 31% of these projects will be canceled before completion. In fact, estimates are that only 16% of software projects in the United States are completed on time and on budget. Even when a project running over budget or schedule was completed, on average only 61% of originally specified features and functions were available in the finished project. The result of cost and time overruns, and the failure of the applications to provide expected features, is extremely costly in terms of lost opportunities, competitiveness, and user satisfaction. Figure 1.1 summarizes some of the published costs for large software projects.

As we consider the usual costs for systems development, we must also keep in mind that development costs are only the "tip of the iceberg." Figure 1.2 displays an example ratio of a project's tasks and the percentage of total costs that can be attributed to each task. The percentage of costs for maintenance (67%) relative to other tasks is quite high. Most maintenance costs are a result of effort expended providing required, but missing functionality, making corrections, and dealing with compatibility issues. Each of these issues relates to requirements, and obviously, developers would like to find ways to reduce these costly expenditures. If these maintenance costs reflected extensions, new features, and integration into other systems, the ratio might be more acceptable. The improved use of maintenance costs is one of the goals for new systems/software engineering processes.

Certainly, the development process has been and continues to be a major issue. Many factors contribute to the problems of delivering computer-based systems. The Standish Group (Johnson, 1995) reports reasons for project failure that include lack of user involvement, lack of executive management support, poorly stated or incomplete requirements, and politics. Individual developers report problems ranging from insufficient memory to cultural shock. Common complaints include inadequate tools, con-

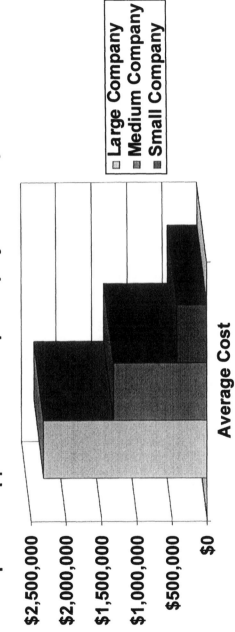

FIG. 1.1. Published costs for software development projects (data from Johnson, 1995).

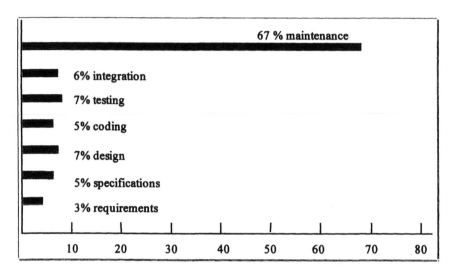

FIG. 1.2. Percentage of an average project's costs throughout its lifecycle.

figuration management nightmares, shifting requirements, and the artistic nature of software engineering.

Problems regularly mentioned by customers include late delivery, invalid information, unfriendly user interfaces, lack of documentation, applications that do not support their needs, and numerous bugs. Commonly heard user complaints include:

- "My software is not upward compatible!"
- "It takes too many keystrokes!"
- "I don't *have* a hand free to type!"
- "The font is too small!"
- "I was promised this three months ago!"
- "This new system *increases* my call handle time!"
- "Why *can't* I put this drawing in this file!"

A BRIEF CHRONOLOGY OF SYSTEMS/SOFTWARE ENGINEERING

There have been many attempts to enhance the management and efficiency of software application development. For years, systems/software engineers have used conventional or traditional approaches such as the waterfall model (Boehm, 1981) for some classes of problems. As projects began addressing knowledge-intensive problems, techniques such as artificial intelligence were used, requiring rapid prototyping approaches

TABLE 1.1.
Characteristics of Traditional/Conventional
Problems and Knowledge-Intensive Problems

Traditional Problems	Knowledge-Intensive Problems
• Calculating rapidly, with high accuracy • Controlling simple, large forces smoothly and with great precision • Storing vast amounts of numeric and alphanumeric data • Manipulating data with great speed • Operating reliably and predictably	• Learning from experience • Perceiving patterns • Dealing effectively with people • Understanding images

to solve development problems (Hayes-Roth, Waterman, & Lenat, 1983). Table 1.1 compares some characteristics of the traditional and knowledge-intensive classes of problems (McGraw & Harbison-Briggs, 1989).

When systems/software development problems escalate in complexity and size, different lifecycle approaches are required. These often include user-driven or user-centered process, architecture-based solutions, iterative development, and quick or rapid prototyping. These relatively new lifecycle approaches are currently being tested against today's development problems.

The transition from traditional approaches to complex systems approaches can be represented historically by some major milestones. Figure 1.3 depicts a transition from traditional code integration, through Structured Analysis and Design Techniques (SADT), into component-based design, and finally to an emphasis on users and their needs. As this historical progression indicates, today there is much more of an emphasis on user involvement and satisfaction in systems development.

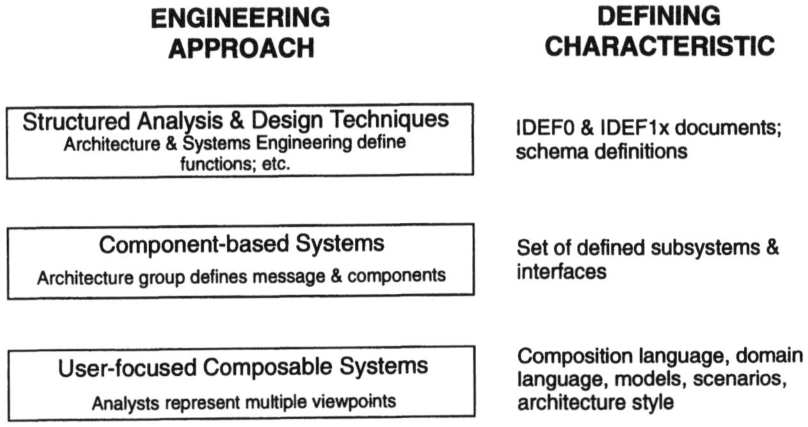

ENGINEERING APPROACH **DEFINING CHARACTERISTIC**

Structured Analysis & Design Techniques Architecture & Systems Engineering define functions; etc.	IDEF0 & IDEF1x documents; schema definitions
Component-based Systems Architecture group defines message & components	Set of defined subsystems & interfaces
User-focused Composable Systems Analysts represent multiple viewpoints	Composition language, domain language, models, scenarios, architecture style

FIG. 1.3. Progression of software engineering.

Systems developers realize they are on the "high velocity" end of the learning curve for engineering large, complex systems. These systems are the target of significant research and development in science and engineering. Solving the problems in large, complex systems development requires the joint effort of investigators in multiple disciplines—computer science, engineering, mathematics, cognitive science, sociology, physics, and information systems to name a few. These problems are also forcing participants in domains and development areas to collaborate on standards, protocols, and definitions.

THE SCENARIO-BASED ENGINEERING PROCESS (SEP)

Some members of the disciplines noted previously, in cooperation with domain performers and system users, have developed an engineering process to address needs in large, complex systems development. Rather than "reinventing the wheel," we have attempted to pull together useful concepts from various development approaches for use in engineering complex systems. These include those shown in Table 1.2. We have concentrated on providing a high-level conceptual process that describes what to do rather than how to do it. Within these "what" guidelines, developers can use their own techniques and favored methodologies, and choose tools and representations that work for them. This overall process has been named the Scenario-based Engineering Process (SEP).

Definition and Applicability

SEP is a user-centered methodology for systems or business process engineering that employs the use of scenarios to scope, bound, and focus

TABLE 1.2.
SEP Reflects Components and Features of Various
Software Engineering Concepts and Approaches

Feature or Component of SEP	Source
• Structured analysis and design	• Conventional systems (development of procedural, well defined, well understood tasks) • Department of Defense 2167a
• Quick prototyping • User-centered systems engineering	• Intelligent or knowledge-based systems (development of well focused, tightly scoped cognitive tasks)
• Domain-specific system architectures	• Complex systems (development of large-scale distributed systems in which multiple viewpoints, changing technology, and hybrid tasks are present)

analysis, design, development, and evaluation activities. Scenarios are used to address both the size issue and the complexity issue. Scenario usage helps scope the development area by setting boundaries around the problem space and the solution space, as shown in Fig. 1.4. Scenarios can reduce complexity by providing a pathway between the specific and the abstract, and by providing a mechanism for decomposition. Finally, and potentially most significantly, scenarios or "war stories" help maintain a connection between domain shareholders and developers throughout the lifecycle. The means by which scenarios accomplish this assistance is explained in the following sections.

SEP is applicable in a well-defined set of situations, such as the descriptors that follow (other situations may call for a more traditional approach):

- *Dynamic environments.* The system must operate in a dynamic environment, or a common dynamic in the environment is changing requirements. Either users are demanding new ways to interact with the system, or the users themselves change. Another common dynamic is the addition of new technologies as new sensors, user interfaces, databases, or telecommunications come on board. Also related to changing requirements is the extension of capabilities or the addition of new features.
- *Human intensive problems.* The system replaces or assists human tasks. Difficulties in automating systems to solve problems of this type include decisions on what part to automate, how to define the human–computer interactions, and how to move back and forth between automation and human tasking. Complexities involve developing reasonable assistance with decision-making, and providing support during situations that stress human capacity, such as "faster than real-time" processing.
- *Long term solutions.* The system or some revision of the system is expected to operate over a significant period. During this time other factors, such as dynamic environment or changing automation (i.e., improvements in available technology), may cause major revisions.
- *Replications.* Partial copies of the application system are needed for similar problems or sites. Thus, a "one-of-a-kind" system will not suffice. It is desirable to reuse as many components as possible rather than start from scratch. Replications may also include extended versions of the original application.
- *Large, complex problem and/or solution spaces.* Size and complexity, in combination with the previous factors, can create a situation calling for a nontraditional approach. In situations where the system may be large but not complex, a simple decomposition approach may suffice.

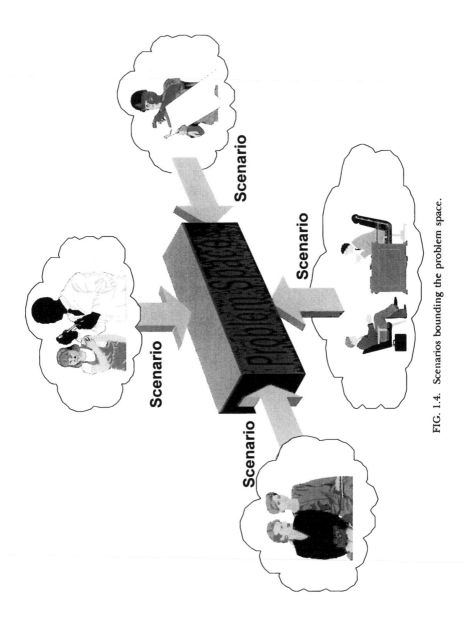

FIG. 1.4. Scenarios bounding the problem space.

SEP can benefit systems automation projects in situations with features such as these because it can provide formalism and structure unavailable in more ad hoc approaches to systems/software engineering.

Characteristics of the Scenario-based Engineering Process

SEP combines features of *business process reengineering* with *systems development* to the benefit of users. For example, SEP focuses on what potential system users or key performers actually *do* in the workplace. It advocates quick prototypes of new processes or systems, which are used to solicit user feedback and to identify issues that need to be explored further. Feedback from users is an explicit requirement in SEP; the process assumes change and enables iteration to evolve and meet new user needs in an affordable fashion. As the new process and/or system is designed and developed, SEP methodologies, techniques, and tools can reduce the cost of automating computer-based systems.

The characteristics shown in Fig. 1.5 epitomize SEP and contribute an aspect to the overall definition of SEP. Together, these characteristics address the situations described in the previous section. Although there are current methodologies with some of these attributes, we believe SEP is one of the few engineering approaches that brings these attributes together in a formal engineering process. The sections that follow discuss each characteristic in more depth.

User-Driven

Users depend on computer-based systems to complete their job tasks more today than ever before. For this reason alone, users should be involved in system development to ensure a seamless human–computer in-

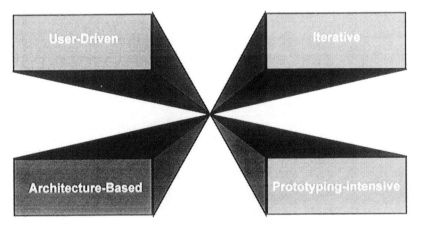

FIG. 1.5. Characteristics or aspects of SEP.

tegration. However, one-time interactions during user acceptance and test will not suffice—users should be involved *throughout* the systems development lifecycle.

User-driven systems development is a challenging prospect, however, and must be planned well, executed carefully, and managed continually. User demands and expectations escalate continually, as they are exposed to better interface features and other technologies. Furthermore, various categories of users often expect different interaction protocols with the system. Yet users demand that the interface remain simple and easy to use, even as the complexity of the system increases. These situations imply that user participation in systems development must be explicitly defined and must be a part of each lifecycle phase. In fact, instead of just *users*, we prefer to use a more general term—*shareholders*—to describe the many, varied interests of users, experts, managers, and others involved in the project.

Shareholders comprise the group of people having a vested interest in the system. Figure 1.6 shows some of the major shareholder categories. In some cases the shareholders comprise a significant portion of the users. In other cases, there are a number of shareholders and a few specific users. Part of the success of any engineering process is identifying the shareholders and obtaining their buy-in. The goals, rationale, and requirements for each category shown in Fig. 1.6 differ, and should be documented by the analysts and developers.

Architecture-Based

Being architecture-based implies an underlying structure or structures for the systems. An architecture-based approach directly addresses long-term solutions and replication situations. Clearly, having a template for

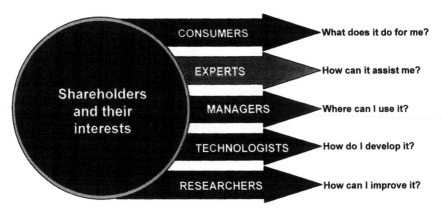

FIG. 1.6. Shareholders in systems development.

replicating systems saves time and money. Also, having a description of the current system, in the form of an architecture, helps developers move from the current system to a revision. Appropriate architectural specifications can even promote the development of a revision. SEP recommends architectures for three major development aspects, including standards, components, and multiple layers.

Tied to the concept of architectures is the use of *standards*. The inclusion of standards in the architecture definition can significantly benefit revision and long-term solutions. CORBA and POSIX are examples of standard architectures at the design level. More abstract architectures can also guide and constrain developers to increase portability, interchangeability, interoperability, and other "-ilities."

Architectures promote the concept of *components*. Components are one means of compartmentalizing the solution and contributing to replication and reuse. A component is a modular piece of the system with defined interfaces. Modularity has longed proved valuable in systems development. Each modular component can be tested individually or integrated with a set of other components for a subsystem test. It may also be reused or replaced during revisions.

Architectures can be specified at *multiple layers* or levels within a system. Significant layers include conceptual (i.e., reference) architectures, site architectures (i.e., architectures for specific sites), application architectures (i.e., an architecture for a specific application), design architectures, subsystem architectures, and implementation architectures. At the design level, architecture definitions include protocols, language decisions, and infrastructure component selections. The application layer architecture, as shown in Fig. 1.7, can be identified by component specifications with interface connections. This definition sometimes is called the "boxes and lines" layer. At the implementation level, architecture definitions can be represented by operating system specifications, macros, parameter typing, and timing specifications.

Conceptual (reference) architectures, as shown in Fig. 1.8, are usually the most abstract layer. They may identify major actors or their responsibilities or activities, constraints on their relationships, and links to requirements.

Iterative

A multidisciplinary systems engineer recently used the history of control theory to provide an excellent analogy for the value of an iterative process. Initially, most problems were solved with a simple control algorithm—a straight execution of code. As the problems became more complex, control theorists transitioned to iterative methods. These methods required that the control algorithm execute for a time sequence with a set of values.

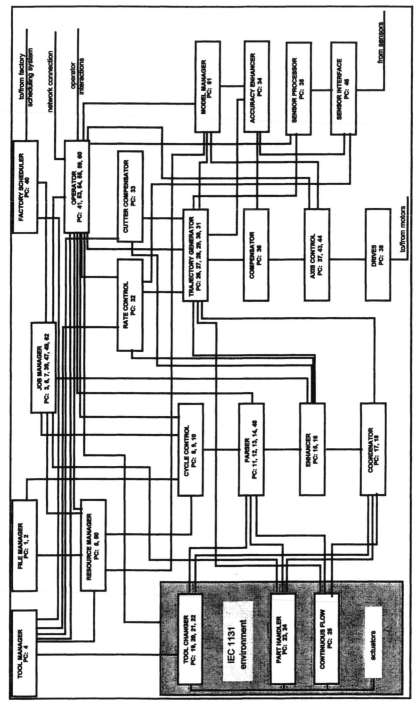

FIG. 1.7. Typical application architecture.

14

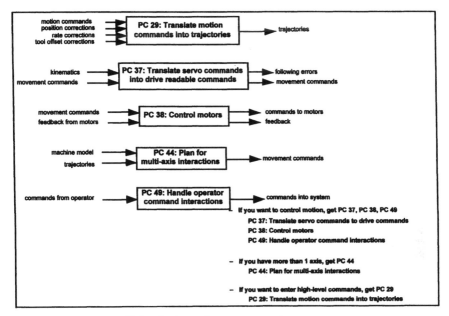

FIG. 1.8. Example reference architecture.

Depending on the outcome, the values were changed for the next time sequence. In other words, the algorithm had to execute and test continually to obtain the desired control outcomes. The same is true of systems/ software engineering in complex situations. The first iteration of systems development may not achieve the desired outcome. The process must allow for subsequent iterations, constantly moving toward the desired outcomes.

Two aspects of iteration are worth noting. The first is the *length of iteration*. An iteration must occur within a timeframe relevant to the users. A product of the iteration may be an animation, a simulation, or other mockup of the system. The important factor is that the iterations occur at a rate adequate for the users.

The second aspect is *formal evaluation*. The engineering process must incorporate evaluation with iterative versions of the system as it matures. SEP emphasizes the inclusion of evaluation from both the perspective of the domain (i.e., is the outcome congruent with the requirements of the domain) and measurement of the development process itself.

The spiral model lifecycle, a variant of the waterfall model, is also an iterative process. However, there is an important difference between lifecycle models like the spiral model and SEP. Spiral models like the one shown in Fig. 1.9 are iterating toward a sense of completion. It is presumed that

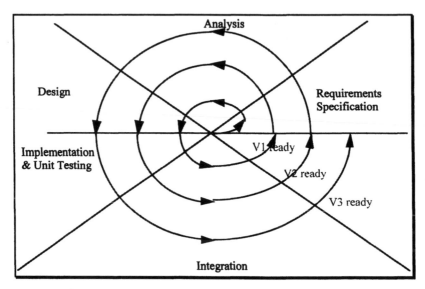

FIG. 1.9. Spiral model lifecycle (adapted from Boehm, B., 1981).

they potentially are capturing *all* the requirements or enabling the development of the *entire* system.

SEP assumes that the system will be changing throughout its lifetime—that there is no single-point "completion." Thus, the iterations may address more specificity at times, or move into new areas or features during iteration. Figure 1.10 illustrates this concept. The first three horizontal bands in the figure (reference activities, application development activities, and technology activities) represent primary SEP activities. Reference activities include developing the reference architecture, domain model, and requirements. Application development activities include the development of the application architecture, models, application requirements, and the application system. Critical technology activities include the development of prototypes and components. Note that these provide immediate feedback to each other, enabling ongoing iteration and refinement. They are not simply spiraling toward a conclusion of the process. Supporting these definition and development activities are engineering process activities and program management activities. Engineering process activities include tools, knowledge acquisition and requirements elicitation techniques, representations, and process guidelines that are provided to developers to enhance productivity, product consistency, and product quality. Finally, program management activities support the entire process. Program management activities are focused on quality and completeness of the process, tasking or assignment of responsibilities, motivation and management of personnel, process facilitation, budgeting, and scheduling.

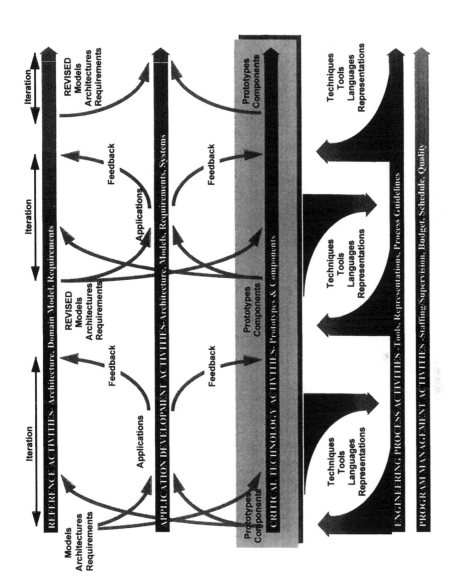

Prototyping-Intensive

One of the major lessons learned in systems development over the last few years is the value of quick prototypes. A prototype is a depiction of some concept to be explained. It may answer a question asked, test the feasibility of an operation, or provide visualization for human interfaces. The field of artificial intelligence (AI) popularized this technique. (Systems/software engineering historically has used feasibility studies or engineering studies in lieu of true prototypes.)

Most AI developers used prototyping as the complete engineering process—that is, without architectures, formalisms, or designs. When prototyping was applied in this manner to a small, focused system that involved one user and one developer, it was often successful. Applied in the right situations, this kind of prototyping can provide invaluable information on performance, as well as a "reality check." It also is very effective for small experiments to test technologies, ideas, or interactions. However, the limitations of quick prototyping appear when it is applied to the exclusion of more formal engineering processes for larger, more complex systems.

Other rapid prototyping approaches transition to a delivery system after some number of iterations. This approach may be successful, but only if the design decisions made for the prototype are also appropriate for the delivery system. Otherwise, performance and extendibility may be affected negatively.

In SEP, prototypes are used extensively; however, more formalism is added to the process via more thorough analysis and documentation. The analysis includes both legacy system analysis and the analysis of user characteristics, work process, job tasks, and decision making. More complete documentation, including knowledge acquisition or elicitation session reports and products, assists developers with component revisions and extensions throughout the iterations. The use of extensive analysis and subsequent formal documentation is incorporated into SEP.

Guiding Principles of the Scenario-based Engineering Process

SEP is essentially a set of guiding principles. Like any set of principles, the implementations may take on many forms, under varying circumstances. We contend that having a set of guiding principles as the engineering process rather than a particular methodology, is important to most system developers. This can help developers avoid believing they have an answer to a system development problem before they completely understand the problem.

For example, the waterfall model discussed earlier has been used extensively for all types of system development problems, even though it is a specific lifecycle process response to a specific type of system development problem. We do not present SEP as a lifecycle process that is the answer to all system development problems. Rather, we present SEP as a set of guiding principles that allows the developers to implement as they judge best.

The four primary principles that have proven useful as we employed SEP are:

1. Focus on engineering products rather than just process.
2. Practice responsibility-based division of labor.
3. Produce distributable engineering artifacts.
4. Employ continuous metric-based evaluation.

Focus on Engineering Products Rather Than Just Process

One of the problems in cost-effective systems/software engineering and business process re-engineering is acquiring the training and experience needed to follow the detailed steps of a lifecycle methodology. This problem includes complying with the steps, and assessing the value of these steps, in producing items that support the engineering process—models, architectures, designs, and prototypes. We advocate taking the approach of "just get the job done." In other words, precisely how you execute the engineering process is not as important as that you do it and produce useful artifacts. For example, a domain model could be produced any number of ways. It is not terribly important how one gets the domain model as long as the constraints (e.g., policies, procedures, production, timeframe of production, resource consumption, and audit trail) are followed. This concept—emphasizing artifacts rather than just process—is in keeping with a model-based approach.

Practice Responsibility-Based Division of Labor

Responsibility-based labor division reduces complexity. Although the concept of *division of labor* is far from new, its execution in engineering processes has not been well defined outside of single-organization, well-understood engineering practices. We contend that SEP helps us apply responsibility-based labor division for the purpose of producing engineering artifacts. Depending on the organizational approach—dictatorial or collaborative—responsibility for delivering the engineering artifacts may be assigned and accepted, or selected.

The authority to carry out the task must accompany the responsibility to complete it. In addition, accountability goes hand-in-hand with acceptance of responsibility. Accountability includes consequences for failure to meet the responsibility.

One reason we emphasize the guiding principle of responsibility acceptance or assignment is the collaborative nature of many current engineering

ing efforts. Many information technology departments collaborate with user/customer groups or third-party consulting organizations. Another example is the numerous large federal programs composed of consortia or loosely federated groups. Without an obvious hierarchy, it is doubly important that each member's responsibility, authority, and accountability be explicitly defined.

Produce Distributable Engineering Artifacts

A model-based approach or process should produce models that can be used and reused. Typically, however, there are serious practical issues that may impede model usage, including nonintegrated tools and incompatible model representations created by different members of the project team.

CASE tool vendors have tried the one-to-one translator approach for interchanging models, but have had limited success on large-scale projects. Another solution for addressing problems with larger scope is the meta model. A meta model contains an abstract representation of all the items in the domain of concern. The meta model is used as the medium to translate to and from various representations and tools. Meta models are one way to avoid the continuous effort to build one-to-one translators among tools and repositories. On the other hand, meta models take time to create and standardize. If more than one meta model exists for a domain, the information represented by different meta models must still be resolved before moving from one tool to another. Figure 1.11 is an example of meta model structure.

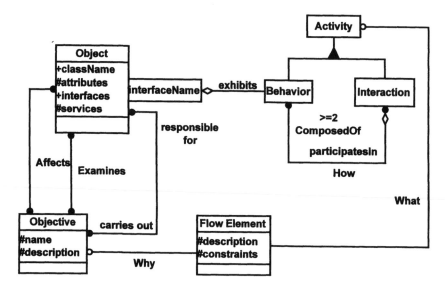

FIG 1.11. Meta model structure example.

Continuous, Metric-Based Evaluation

Historically, many software development projects have been exempt from the rigorous evaluation and testing that most science and engineering projects undergo. In today's environment of cost-effective quality management measures, evaluation is becoming an integral part of any project. We advocate that evaluation take place continuously and proactively during development. By proactive we mean that evaluation must be defined as a significant part of the process from the beginning of the project.

As we discuss in more depth in chapter 12, there are two major areas of evaluation. One is the measurement of outcomes from the domain perspective. Does the installed system actually produce changes in the situation? The second evaluation area is the performance of the system itself. The first of these is very domain intensive, and the second is technology intensive.

A baseline is an essential component of the evaluation. Metrics and their baseline values must be established prior to the evaluation. Working from the baseline, we can measure outcomes produced by iterations of the system solution. During the iterations, the system solution may take the form of prototypes, simulations, testbeds, and demonstrations.

Another requirement for the application of the final guiding principle for SEP is a goal or mission, which may take the form of project "road maps," scenarios, or vision statements. We break down the goals to define the desired objectives that, when met, will ensure that we have met the goal. Our evaluation is then focused on assessing the extent to which we have achieved our objectives.

SEP—A MODEL-BASED SYSTEMS DEVELOPMENT PROCESS

Following SEP principles, analysts use multiple knowledge acquisition and elicitation techniques to understand the domain, its users, and its requirements.[1] The output of this activity enables analysts to envision "to be" scenarios that represent better ways of working, or efficient application of systems to reduce problems revealed during elicitation sessions. Models are developed to represent the domain, and components are defined that would support its preliminary requirements. The new or envisioned scenarios are brought to life through prototyping. As these prototypes are demonstrated or exercised, shareholders evaluate them to help developers refine and extend the models and requirements. Once the models and

---

[1]Suggested techniques are presented in chapters 4 through 11.

requirements are better understood, developers use previously constructed architectures, models, and components as a springboard to the development of the new application.

Traditional views of lifecycle approaches discussed previously are process-focused. The summary view of SEP presented in the preceding paragraph presents a model-based, product-focused view of a lifecycle. The strength of a model-based view is the visibility of the feedforward and feedback loops between the systems and the models. This view explicitly depicts the ability of SEP to help developers create new versions of systems from the existing models and then to update these models based on the development and acceptance of systems.

Models are the underlying structure that enables SEP engineers to create architectures, build prototypes, iterate over time, and maintain interaction with the users. In essence, models are a major mechanism for reuse capability. The section that follows describes types of models used in SEP.

Types of Models in SEP

Primary types of models include those that follow, which are described in more detail in the subsequent paragraphs:

- Domain model.
- User models.
- Site models.
- System models.
- Performance models.

Domain Model

A *domain model* describes the domain of concern. This model includes information on the primary concepts, objects, and entities in the domain. This typically includes information such as actions and behaviors, attributes, relationships, and contextual information.

User Model

User models are actually a subset of the entire domain model. The user model represents a user profile, which includes information on user characteristics such as computer expertise, domain expertise, preferences, responsibilities, work processes for which they are responsible, and requirements for performance.

Site Model

The *site model* also is a subset of the entire domain model. It represents detailed, specific information on a particular location or site that exists within the domain. For example, a Clinical Associate system would have to operate in numerous different types of clinics. Part of the modeling activity is defining the baseline domain—one selected clinic. The site model involves the development of a description or model for a particular type or "instance of" a clinic, such as a dermatology clinic, a urology clinic, and so on. Although site models are similar to each other in structure, key differences (and thus, requirements) among sites should be reflected in their models.

System and Performance Models

As the lifecycle process for a system moves to the point of development of a specific application, the project team references the system models, performance model, and other models developed earlier. A *system model* represents a generic version of a system and identifies the basic components and activities required for a particular type of system. A *performance model* represents possible performance constraints and requirements for the domain. These models become default guidelines or templates for the development of specific application systems.

Figure 1.12 shows various models contributing to the design and development of Clinical Associate systems for health care (e.g., domain models, performance models, user models). As each version of the system is prototyped and its components developed, more is learned about the domain, performance requirements, and user needs. The result of evaluation of the different versions of the Clinical Associate system feeds back into the models to refine them.

There is a significant dependency between SEP models and architectures. A set of architectures exists, matching and building on the blueprints provided by the models—the reference architecture (which is domain relevant), site architectures (which are site relevant), system architectures, system designs, and implementation profiles.

If we expand the simplistic model-based systems development view in Fig. 1.12 to consider the sources of information for these models, we begin to see how critical knowledge acquisition and requirements elicitation activities are to the entire process. As Fig. 1.13 illustrates, initial models are built from information elicited and/or analyzed from sources such as experts, users, standard documents, policy documents, recipients, existing models, and other shareholders. Using this information as a foundation,

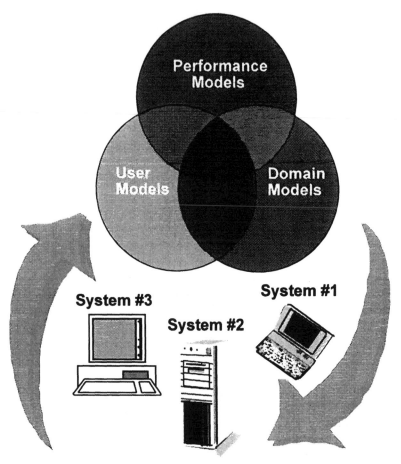

FIG. 1.12. Model-based systems development is a characteristic of SEP.

analysts and engineers attempt to build exemplar (i.e., default or guiding) models. Exemplar models help us guide and manage iterations within a large, complex systems development project.

Throughout this process, scenarios help define the domain and its work processes and requirements. Analysts and engineers decompose these scenarios into work processes and tasks, which are used to develop performance models, ontologies, and standards (see chapter 2). This information is documented, using procedures, forms, templates, and representations defined for the project (see chapter 3). The documentation ensures traceability and easy access and use during the rest of the development cycle.

Using the scenarios, products of the analysis process, and models, a system architecture is developed to support the models and provide blueprints for component development. Additionally, the required components

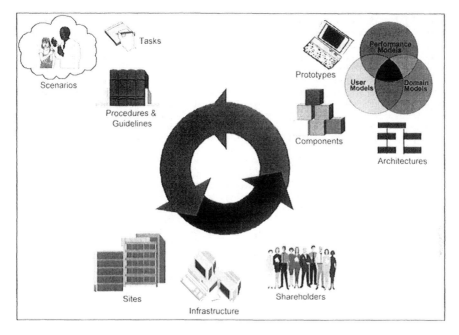

FIG. 1.13. A context for model-based development illustrating the role of information sources and analysis activities in building and refining models.

for the system are identified. Prototypes are developed for the purpose of refining the requirements and designing the required components.

Models acquired and represented during knowledge acquisition, requirements elicitation, and follow-on analyses become part of the domain model. Information from the domain models is used during development for analyzing performance models, for composing prototypes and components, for defining user interfaces, to visualize slices of context and perspective, and as a basis for the architectures.

BENEFITS OF THE SCENARIO-BASED
ENGINEERING PROCESS

SEP provides developers with at least two benefits not found in other lifecycles. First, it enables developers to bring together and benefit from the strengths and tools of both systems/software engineering *and* business process re-engineering. For example, SEP includes modeling, formal representations, analysis and design techniques, and prototyping—techniques and tools from systems/software engineering. SEP also is a user-centered process, with the ability to represent business enterprises, tasks, and performer decision making, and the redefinition and refinement of work processes—techniques and tools borrowed from business process re-engineering. This combination of techniques and tools from formerly disparate worlds, and the synergy that results from it, is one of the primary strengths of SEP.

The use of scenarios, the central tenet of SEP, is an example of this synergy. Traditional systems/software engineering uses cases primarily for testing, which occurs at the end of the development cycle. Business process re-engineering indirectly uses example cases or "war stories" as sources for process descriptions, which occur during the initial phases of the re-engineering process. SEP expands the concept of "use cases" to include the war stories and episodic knowledge (as it is viewed in business process re-engineering) and employs them throughout the system development process. These scenarios provide the following benefits:

- A common forum for building a relationship with the users or experts.
- A means of scoping and bounding the problem space to be discovered.
- A means of prioritizing the potential solutions from a performance perspective.
- A focus mechanism for scheduling areas to attack.
- A context descriptor for diagrams and models.
- The conceptual level script for demonstrations and system use.
- A product in the transition from problem space to solution space.

The second primary benefit of SEP is that it provides assistance in solving one of the most difficult problems facing developers and users of complex systems—transitioning from the problem space (i.e., the domain) to the solution space (i.e., automation). This process requires a smooth and meaningful transition from the knowledge, requirements, and various models existing in the domain of concern to the solutions (e.g., computers, equipment, human–computer interactions, and other products). SEP provides processes and products to support making this transition and documenting an audit trail. SEP takes some of the mystery out of the transition process by making the activities explicit and the goals attainable.

SUMMARY

The systems development process continues to be a significant problem when developers attempt to build solutions for large, complex problems. Historical lifecycle processes are inadequate in addressing these types of problems. We have described the Scenario-based Engineering Process (SEP), a process that combines strengths of both traditional systems/software engineering and business process re-engineering.

The four characteristics that help define SEP were described as follows: user driven, architecture based, iterative, and rapid prototyping. These are the underpinnings of successful engineering processes that address large, complex problems. These four characteristics are also well-recognized attributes of both good systems/software engineering and business process re-engineering practices.

In addition to these characteristics, SEP is based on four guiding principles. The first of these, *focusing on engineering products,* demands that project teams emphasize the products rather than follow lockstep in line with a process. This requires that the creator of the products is proactive, is motivated, can identify necessary resources, can recognize goals, and employs appropriate constraints and guidelines.

The second principle, *practicing responsibility-based division of labor,* also is relevant to engineering products or artifacts. It requires that the manager assigns, or lets performers select, responsibilities that they are accountable for completing. The third principle is *producing distributable engineering artifacts.* This principle implies the use of meta models—abstract representations of the domain—that can be used and reused.

The final principle is that SEP *enables continuous, metric-based evaluation.* This includes measuring outcomes or impact of the system on the domain and an evaluation that determines the performance of the system. To put this principle into practice, developers must determine baselines, set goals, and define objectives for the evaluation.

SEP has at least two primary benefits for systems development. First, it integrates the strengths from two engineering practices—systems/software engineering and business process engineering—into an "umbrella" lifecycle process. However, the process is not dictatorial. Developers select from the transition processes and products those that work for them in their organization's lifecycles. In other words, we composed an umbrella process and defined its products in a way that allows selection of pieces attractive to various systems developers. Secondly, SEP provides a defined transition path from the problem space in the domain of concern into the solution space of semiautomation. Along the way, it provides guidance for transitioning from the problem to the solution and maintaining contact with the shareholders.

The chapters that follow expand the description of SEP and how it is practiced. In chapter 2 we discuss primary SEP activities and the models and other artifacts that are produced during these activities. In chapter 3 we discuss how to put SEP in practice, including tips on planning and managing effective requirements or knowledge acquisition activities. In chapters 4 through 11 we present specific requirements elicitation or knowledge acquisition and analysis techniques that serve as the primary vehicle to help developers apply SEP activities. In chapter 12 we present an evaluation methodology that is congruent with SEP goals and approach.

Chapter 2

Engineering Activities and Artifacts

In chapter 1 we presented a general description of the Scenario-based Engineering Process and the rationale underlying its use. One of the guiding principles of the SEP is to "focus on engineering products rather than just process." Throughout the SEP, development teams produce a number of different types of products as they interact with users and other shareholders, observe work process, investigate legacy systems, facilitate group sessions, and develop models and architectures.

In this chapter we provide more detail on the engineering activity areas that comprise the SEP and the artifacts produced by each activity. These details demonstrate the efficacy of the two primary benefits of the SEP (chapter 1) and prepare the reader to plan and conduct the required knowledge acquisition and elicitation sessions.

First, we briefly discuss the roles of knowledge acquisition/elicitation and analysis in the SEP. This includes examining the "front end" process for knowledge acquisition and requirements elicitation, and introducing the initial and intermediate SEP artifacts or products that are produced. Intermediate products—scenarios and task descriptions—are used during each of the SEP engineering activities. We then define and explain the use of the critical products produced during each of the major activities of SEP—reference activities, site or application activities, and component or technology activities. This includes describing the purpose of each product or artifact and how it is represented. Throughout the chapter, we refer to subsequent chapters that explain how the knowledge acquisition or elicitation program should be set up and managed, and how to select and use elicitation and analysis techniques to ensure the production of usable artifacts.

This chapter builds on the basic information about the SEP presented in chapter 1. Its purpose is twofold: (a) to enable analysts and engineers to understand the relationship among the engineering activities and the critical artifacts produced by each, and (b) to recognize the role of documentation, knowledge acquisition, and analysis in the process. Specifically, the chapter is designed to help the analyst and engineer achieve the following goals:

- Identify the major engineering activities within the Scenario-based Engineering Process.
- Recognize the importance of documentation and analysis in the process.
- Recognize that knowledge acquisition and elicitation techniques are the tools used to produce the engineering artifacts.
- Define the critical products produced in each of the major engineering activities of SEP.
- Identify the uses and representation of the critical products produced by engineering activities.

THE ROLE OF KNOWLEDGE ACQUISITION, ELICITATION, AND ANALYSIS

In the not-too-distant past, requirements engineering focused primarily on understanding data format, input/output, and use. Users rarely were consulted when developing these data-driven systems. Screens were designed based on the "best fit" for the data that needed to be displayed. In this environment, developers used techniques that focused on collecting data from existing system documentation and forms, and analyzing that data to define requirements and design.

Knowledge Acquisition and Requirements Elicitation

One of the primary characteristics of the Scenario-based Engineering Process (and other user-centered approaches) is that it requires much more of developers than did older methodologies. Developers using the SEP still must be able to analyze data from legacy systems and artifacts, because the new components and processes often must fit within existing system environments. In addition, however, today's developers must also be able to conduct extensive domain investigation and analysis, which requires considerable interaction with shareholders and users. Figure 2.1 illustrates the knowledge acquisition, requirements elicitation, and analysis activities that occur using the SEP.

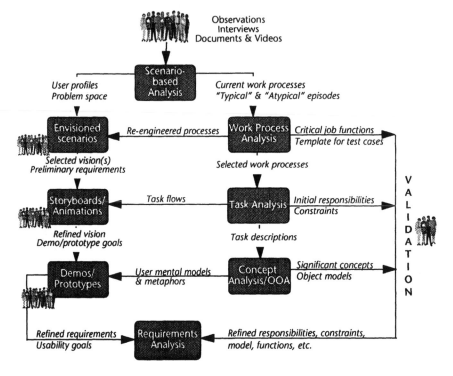

FIG. 2.1. SEP requires more extensive knowledge acquisition, requirements elicitation, and analysis activities in what would normally be considered the requirements phase of a project.

SEP starts with knowledge acquisition and elicitation activities involving selected users in the domain. Scenarios that describe the domain and the requirements of the users' work processes are elicited and documented. Important characteristics of the users are identified and used to compile a user profile including information such as domain expertise, computer expertise, experience with graphical user interfaces, age, and other factors that may impact usability requirements. The outputs of scenario analysis include user profile definition, an initial "problem space" for the system (represented as scenarios), an understanding of current work processes, and both typical and atypical episodes or scenarios.

Next, work processes are examined to select those that will be analyzed further and to identify opportunities for refinement through automation or process revision. The output of the work process analysis includes re-engineered work processes, work processes selected for further analysis, and a template for future test case creation. The re-engineered work processes are used to envision how the new system or processes might work.

Users participate in the development of envisioned scenarios and in a subsequent review of these scenarios to select those critical enough and representative enough to be used as the framework for demonstration or prototype systems. These envisioned scenarios become graphical and narrative "preliminary requirements."

Analysts are also using the selected work processes as input for task analysis. The goal here is to understand better what work must be done to successfully operate in the scenarios and work processes already described. As a result of task analysis, task flows are produced that provide a graphical breakdown of the major activities in the task. In addition, task descriptions are produced that provide detailed information on the performance of a task. The result of these activities is a better understanding of the initial responsibilities and constraints for the new system, as well as information that can be used to develop demonstration or prototype storyboards and screens.

Users and other shareholders participate in the development, review, and refinement of storyboards and animations that represent the selected vision and the task flows. During this process, the envisioned scenarios may be refined. This activity provides the project team with a better understanding of the type of scenarios the system must support, and goals for any demonstration or prototype system that will be developed.

Concept analysis and object-oriented analysis is also being conducted, operating on the task descriptions. Significant concepts that are represented and used throughout the task descriptions or that are critical for the successful completion of tasks are identified. Analyzing these concepts helps developers understand the mental models used by shareholders in the new system. Concept analysis also may provide clues, such as metaphors for work, that can be used to structure the new system and develop usable icons for buttons and other screen elements.

As the demonstration or prototype system is developed, users and other shareholders again participate in screen development and refinement and in reviews of the resulting product. This input provides critical information that can help developers refine the initial requirements for functionality and usability.

To meet the goals of user-centered systems, and to enable processes such as that presented in Fig. 2.1, development teams now must be able to elicit information about, and analyze, the following:

- Knowledge (e.g., conceptual) about the domain.
- Information about the characteristics of the users.
- Details about the work processes and primary tasks, including how information is used to complete them.
- Decision heuristics that are used to work "smarter."
- Problems that impede the current processes.

- Critical success factors and/or return on investment (ROI) goals that must be considered in redefining the way work is done (e.g., reducing training time, reducing errors, etc.).
- Opportunities to refine work processes and tasks through the definition of new processes or system automation.

Carefully selected and appropriately used, knowledge acquisition or elicitation techniques become the mechanism that enables developers to achieve the goal of gathering, analyzing, and using the information that user-centered processes require. Some development teams believe they can successfully achieve these goals simply by interviewing users and conducting surveys and/or questionnaires. As noted in chapter 7, however, these techniques limit the respondent to surface-level information which is not necessarily accurate, complete, or at the depth desired. Techniques (chapter 4) such as work process and task analysis, concept analysis, decision process tracing, and focus groups, are extremely helpful in ensuring that requirements address not just surface features, but semantic features of a domain. Systems defined and developed using the SEP (and the techniques we describe in later chapters of this text) meet the needs of users because they take into account the following:

- How a business process or task should be re-engineered.
- How the system should support the users' tasks and decision making.
- How components should function to ensure congruency with the user's mental model of the domain and work.
- How the system should present itself to the users.

These requirements of a user-centered approach mean that development teams must carefully select and employ knowledge acquisition and elicitation techniques and spend more time in the initial phases of development than before. Chapter 3 describes how to set up and manage these efforts. Chapters 4 through 11 present information about selecting and using the appropriate techniques.

Documentation Products

In addition to eliciting the requirements-related information, development teams should carefully document their findings and then conduct the analysis required to extract meaning and products from session output. Initial output or products of knowledge acquisition or elicitation sessions with users, expert performers, and other shareholders include the documentation of these sessions. A number of different types of sessions (see chapter 4) may be conducted during the SEP, including:

- Interviews
- Observations
- Scenario analysis
- Work process or task analysis
- Decision process tracing
- Group sessions (e.g., focus groups, brainstorming, etc.)
- Document analysis
- Other types, including legacy system analysis

After the analyst conducts a session, we suggest that he or she write up a report of the session, using a session report template (see chapter 3). We refer to this type of documentation as knowledge acquisition or requirements elicitation *session reports.* Session reports contain the following information useful to members of the development team:

- Name of the analyst conducting the session.
- Name of the user, expert performer, or shareholder participating in the session.
- Type of session, or technique(s) used.
- Statement of the session goal and objectives.
- Summary of topics discussed and findings.

The session report also provides traceability for an idea or piece of information back to its source. We suggest that initial documentation such as the session report be stored in such a way that it can be accessed and reviewed by all team members.

Other types of initial documentation include audiotapes or videotapes of sessions, videotaped observations and walkthroughs of a process within the domain. Existing training and system documentation also may provide useful initial information.

The information contained in the documentation is used continually throughout the SEP. Engineers and analysts may review information previously captured, refine initial understandings of information, and seek a source for a particular piece of information. In addition, these reports become the fodder from which engineers conduct analysis and produce important SEP artifacts such as models, requirements, and architecture

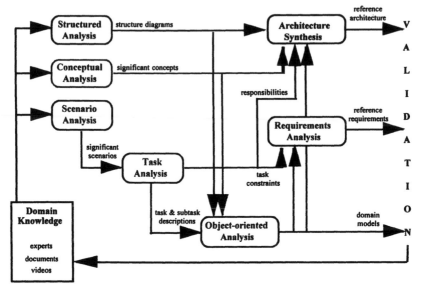

FIG. 2.2. A useful analysis process in the SEP.

specifications. The next section describes a useful analysis process in the SEP and the types of products or artifacts typically produced.[1]

Analysis Process

The SEP requires that artifacts or products be produced to help developers transition through iterations of the system process. However, SEP does not force developers into predetermined process steps. Over time, we have discovered some processes that have been effective on our programs. We present them here as examples, not requirements, for developers.

In situations in which the developer must accommodate legacy systems (i.e., ones currently in operation), quickly produce prototypes, and create reference level artifacts for future use, the analysis process and techniques shown in Fig. 2.2 may be useful.

The structured analysis technique is applied to the legacy system designs and descriptions. Conceptual analysis is used to structure and document the domain familiarization process and to identify significant concepts. Scenario analysis is used to identify significant scenarios, which help communicate the functional requirements of a system. Using task analysis, analysts decompose the significant scenarios into the primary tasks, subtasks, and task requirements (e.g., constraints, context, information, re-

---

[1]More extensive information on the various types of analyses typically used in SEP can be found in the individual chapters detailing how to use a particular technique.

sponsibilities, resources, etc.). Using information from the task analysis, developers make a transition to an object-oriented paradigm and produce domain models. Additionally, task constraints are fed into the requirements analysis to contribute to the development of reference requirements. Responsibilities identified during task analysis are fed into the architecture synthesis process. The feedback and feedforward arrows display the movement of the artifacts or products throughout analysis. These same techniques are used throughout the iterations of the SEP, and the resulting artifacts or products are refined.

This analysis process must be scoped and focused. For example, it would not be feasible to complete an in-depth analysis of every scenario described, nor to conduct an in-depth task analysis of every task performed by each shareholder in a domain, or at a site. Instead, only significant scenarios should be analyzed. Complete task analyses should be conducted only on high-priority tasks. Figure 2.3 illustrates the task prioritization process. This determination is based on the significance of the concepts represented in the task, the significance of the scenarios from which the tasks were derived, analysis of existing structure diagrams, and a review of existing process descriptions.

Once high priority tasks have been identified, task analysis techniques are used to define each priority task and its subtasks, identify task constraints and responsibilities, and investigate information and resource requirements.

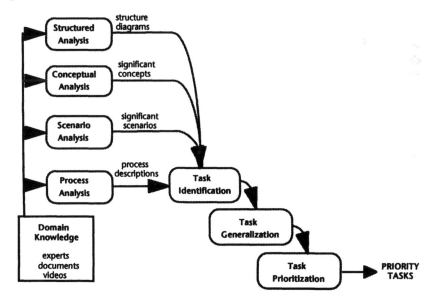

FIG. 2.3. A process for prioritizing and selecting tasks for further analysis.

ACTIVITIES AND ARTIFACTS WITHIN THE SEP

The initial output of the SEP includes session reports and other basic types of documentation. As these products are used and analyzed, and as other engineering activities occur, additional products and artifacts are produced and refined.[2] These include the intermediate products, which are used throughout all of the SEP activities, and special products resulting from the specific SEP activities. Intermediate products include scenarios and task descriptions. Using these intermediate products, engineers and analysts produce products and other artifacts (e.g., models, requirements, architectures) as suggested by Fig. 2.4.

Each SEP activity produces models, requirements, and architectural products. Models that are produced include domain models, site or application models, technology models, performance models, and computational models. Various requirements that may be produced include reference requirements, site or application requirements, technology requirements, and performance requirements. Architectural products that are produced include reference architectures, site or application architectures, designs and component specifications, implementation profiles, and standards.

The following section describes the types of intermediate products that are created within the SEP. If the resulting models, requirements, and architectures are to be well-defined, the accuracy and completeness of these intermediate products are of critical importance.

Intermediate Products and Artifacts

During the elicitation, acquisition, and representation of information and knowledge, a number of products are created, such as the documentation and videotapes described earlier in this chapter. From these, engineers and analysts produce intermediate products, including scenarios and task descriptions, that may be supported by graphical depiction of domain processes. These products are refined during many subsequent engineering activities. They are used to help engineers and analysts make the transition from the concrete to the more abstract, formal representations that eventually will be needed. The sections that follow define the critical intermediate products, their use, and some representation options.

Scenarios

A scenario is a specific instance, case, experience, "war story," or example that happens over time. A scenario contains a description of the environment, the context, the actors, and the actions. It has definite beginning and end points. The rationale supporting inclusion of the scenario is attached to the scenario description.

---

[2]Evolutions of each artifact must be controlled through configuration management.

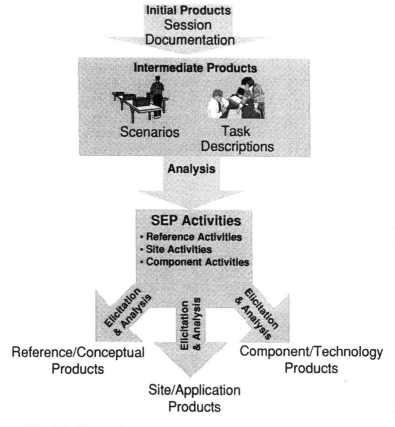

FIG. 2.4. The product production process for the SEP is driven by knowledge acquisition and elicitation sessions, analysis, and the primary SEP engineering activities.

Using Scenarios. Scenarios serve a number of purposes in the SEP, as initially presented in Fig. 2.1. First, they are an easy way to express shareholder requirements. Second, they promote an ongoing relationship with the users or experts during the engineering process. Third, they can represent a user's vision. Fourth, the scenarios are the goals against which we measure prototype design and development progress (see chapter 12). Fifth, scenarios help engineers scope and bound the problem space and the solution space. (This focusing includes prioritizing the problem areas or potential solutions to be addressed.) Sixth, scenarios can become the conceptual description for a demonstration or prototype system. Most important, the scenarios contain the *context* that make domain models meaningful and grounded. Throughout the SEP, developers should maintain a link between the scenarios and the items derived from the scenario.

Representing Scenarios. Scenario analysis is the knowledge acquisition technique used to acquire and document the scenario information (see chapter 5). Once captured, scenarios may be represented through video clips, drawings, timelines, text outlines, simulations, animations, event traces, and process diagrams. The temporal visualizations (moving or animated) are usually most meaningful to the shareholders and are thus easiest to validate for accuracy and completeness. On the other hand, text outlines and diagrams are the most amenable to automated translation.

Figure 2.5 is an example of the simplest representation of a scenario—a text description. Working from this description, we could begin to analyze what happens to each of the victims, which emergency medical personnel interacts with each victim, what information is captured and generated, what resources are used, what treatment each victim receives, and so on.

Another type of representation is to view the scenario against a timeline, as shown in Fig. 2.6. In this case, the line across the top represents the timeline, from 0 to 15 minutes after the start of the scenario. With this representation we can see not only the main activities (assess, treat, etc.), we can also determine which resource (EMT–P) completed the activity, which patient (P7, P10, etc.) received the action, and the point in time the activities occurred.

Chapter 5 illustrates other techniques that can be used to document and analyze a scenario, including topologies or "maps," event traces, scenario matrices and models, and flowcharts. Commonly used representation techniques such as process diagrams and activity state diagrams also can be appropriate mechanisms to represent scenarios. For example, Fig. 2.7 illustrates a process diagram that was developed to depict a scenario. In this diagram, processes are illustrated to communicate where the process occurs, and how the location-specific processes feed into the next site or "scene."

- Incident at bus stop; folks lined up along sidewalk
- Taxi driver on PCP jumps curb and mows down people
- Four children and seven adults involved
- First call out describes event as "car accident"
- One ambulance with two Emergency Medical Technicians–Paramedics (EMT–P) arrives
- They begin triage
- EMT–Ps called for other emergency components
- Over time another ambulance with 4 EMT–Ps, 6 EMTs, and 12 firefighters (FFs) arrive
- Eventually, the call involves 5 ambulances and 2 full engine companies
- Triage results in 1 presumed dead adult, 2 children in cardiac arrest
- All patients sent to D.C. General

FIG. 2.5. Textual description of a scenario, provided by RN.

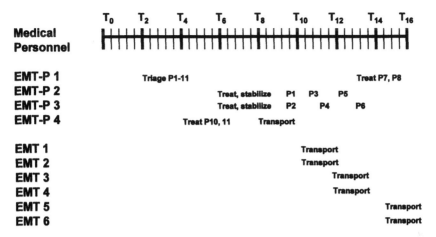

FIG. 2.6. Timeline view of scenario (P = patient; T = time count/minutes).

One of the most useful representations has been the interaction diagram, as illustrated in Fig. 2.8. The interaction diagram is valuable because it is capable of representing *both* data and processes for a scenario and maintaining the connections between the two. Interactions diagrams also have been well accepted by user communities. Additionally, it is also fairly easy to map them to tasks and the meta model (i.e., the model of the domain model, including terms and structure).

Task Descriptions

A task is a significant, defined, bounded, meaningful event in the domain.[3] Sets of tasks, examined as a whole unit, become the work processes that define a job. Tasks are derived from scenarios and represented via numerous variations of task description mechanisms. Task analysis involves examining and analyzing a task into its components parts to define cues to initiate it, subtasks, information and resources used, constraints, and other required descriptions.

Using Task Descriptions. Experienced shareholders can isolate and identify the most significant tasks within scenarios. Once a task is identified as a significant, meaningful event, task analysis is used to identify required information about the task. This may involve the use of knowledge acquisition or elicitation techniques such as structured interview, timeline analysis, and functional analysis. (Chapter 8 describes procedures for conducting task analysis and documenting the results.)

---

[3]Chapter 8 provdes a more in-depth definition and background information.

Mortar fire with injuries scenario

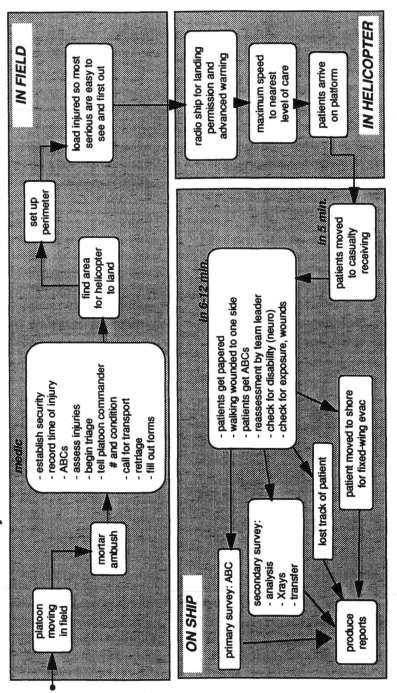

FIG. 2.7. Process diagram for a scenario.

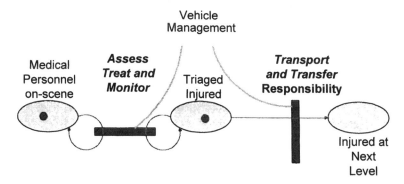

FIG. 2.8. Interaction diagram representing a scenario.

As was illustrated earlier in Fig. 2.1 and is shown in Fig. 2.9, task descriptions are critical in enabling the transition from domain scenarios to technology development. This transition requires the use of two representation languages—a domain description language and an architecture language. A domain description language is a semantically rich language with specific terms for the domain of concern. An architecture description language is a formal language for representing architectural items, such as interfaces.

Representation of Task Descriptions. Task analysis breaks down tasks into descriptions that can be documented via representations such as the task frame, as illustrated in Fig. 2.10. The task frames may be maintained in a

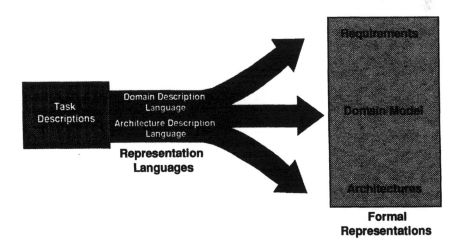

FIG. 2.9. Using task descriptions to begin the transition to formal representations.

Task overview:	The provider recognizes the patient is ready, takes the Medical Record from the bin, scans the relevant information, mentally prepares for the exam, and escorts patient into the exam room
Performer(s) involved, and responsibilities:	The Healthcare Provider is responsible for this task
Cue to initiate task:	Recognition that Medical Record is in the bin (signalling that patient is ready)
Temporal requirements:	Seconds
Subtasks:	1. Recognize patient is ready
	2. Retrives Medical Record from the bin
	3. Scans information on Form 600
	4. Prepares mentally for the exam
	5. Walks to the waiting room
	6. Calls patient
	7. Escorts patient to exam room
Location:	Exam area
	Waiting room
Information requirements:	Patient name (current Form 600)
	Patient complaint(s) (current Form 600)
	Patient vitals (current Form 600)
	Patient general information (current Form 600)
	Relevant patient history (from previous Form 600)
Resources used:	Medical Record
(and source)	Current Form 600s
Constraints:	N/A

FIG. 2.10. Sample task frame for representing task descriptions.

relational database system. The fields in the electronic records can be parsed to create formal model structures such as object models, entity-relationship diagrams, and process diagrams.

The task descriptions within these frames are discrete fields with values. The values of the fields include semantic terms in the domain. The fields on the frame were selected because they capture important information required by the meta model. Thus, the task frame fields also represent the structure of the meta model. When the fields are filled with values for a task, that task information and the values can be converted to a formal domain model.

The information within a task frame is also used to compose the conceptual reference architecture. To contribute to the conceptual reference architecture, tasks must first be decomposed into subtasks. (At some point, based on criteria defined by the project, it does not make sense to decompose further.) These elemental tasks become *responsibilities* in the reference architecture.

If tasks are decomposed iteratively into even greater specificity, they may become functional procedures in the computational model. This decomposition results in a transition from domain-relevant tasks to technology/computational-relevant tasks, and results in a procedural design.

Process coordination diagrams may also be used to represent hierarchical decomposition, specializations of tasks, and the coordination of tasks, as Fig. 2.11 illustrates. Resources involved in the tasks can be listed along the arcs. By using a different type of arc, the engineer developing the task coordination diagram can differentiate between decompositions as shown in this figure, and specializations of tasks.

The most critical intermediate products—scenarios and task descriptions—are used throughout all subsequent SEP activities. Analysts and engineers use the intermediate products as input for the SEP activities. Using the knowledge acquisition and elicitation techniques described later in this text, analysts and engineers perform the following activities and produce additional products, as shown in Fig. 2.12:

- Reference or conceptual activity;
- Site or application activity;
- Technology or component activity.

Reference or conceptual activities produce knowledge and products that feed site or application activities, which in turn produce knowledge and products that feed technology or component activities. Specific site or application activities may reveal information that should be used to refine the reference models, requirements, and architecture. Products typically produced during each activity appear outside the appropriate activity band. The sections that follow discuss these products in more detail. Under-

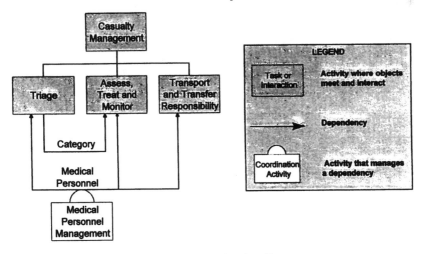

FIG. 2.11. Task coordination diagram.

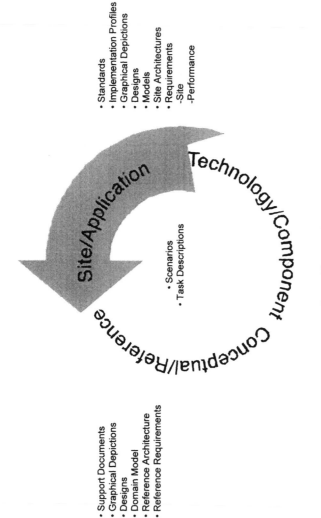

FIG. 2.12. Each SEP activity area produces products and artifacts, using scenarios, task descriptions, and input from previous SEP activity phases.

standing this information enables analysts to plan knowledge acquisition and elicitation sessions better, select appropriate techniques, and develop more usable products.

Reference or Conceptual Activity Artifacts

The reference or conceptual level activities are an abstraction from the actual sites or application areas being addressed. The idea at the reference level is to provide a representation that can cover many sites and many application systems. The reference level is achieved through generalization, specification, and synthesis. The primary products or artifacts of the reference/conceptual level activities within the SEP include the domain model, reference architecture, and reference requirements.

The reference activity artifacts should represent both current and future situations and systems (i.e., "as-is" and "to-be" [or envisioned] models and architectures). Much of the information needed to establish these models, architectures, and requirements is derived from the site and technology activities. The information yielded from site and technology activities, plus additional information acquired from general coverage or conceptual activities, is used to compose the reference models, architectures, and requirements.

Domain Model

The domain model is the *ontology*, including structure, semantics, and syntax, of the domain of concern. By ontology, we mean the entities, their activities, their objectives, and their relationships which, used together, define the domain. (This information should be organized in a formal representation so that it is accessible in electronic form.)

Using Domain Models. The domain model is created and maintained based on input from domain shareholders. During the development of systems, domain model information is used to populate databases, define components, and depict coverage. Systems may access domain models during execution.[4]

Representing Domain Models. The domain model itself is usually coded in a representation language. For example, many languages and tools could be used to represent the domain model. A viewer of the domain model would see lines of code written in the programming language. However, this form is not necessarily useful to knowledge acquisition or domain experts. Nonprogrammers involved in analysis or validation usually prefer some other mechanism than formal representation language code.

---

[4]Configuration management of domain models is a real concern for their active use.

Thus, for the most part, the domain model is viewed in a graphical form. A graphical view of the domain model is only meaningful if the view includes labels defining its perspective and context. If an arbitrary graphical view is presented without a perspective and context, the result is usually an argument about its correctness.

Graphical representations for the domain model include object models, conceptual models, process diagrams, entity-relationship diagrams, interaction diagrams, event trace diagrams, modified flow diagrams, and others commonly used in knowledge acquisition, elicitation, and engineering. A sample object diagram and a process diagram are shown in Fig. 2.13 and Fig. 2.14, respectively. Other representations will be presented throughout the text in descriptions of technique use and output.

Reference Architecture

The reference architecture is the conceptual framework for the domain and is used to compose systems. This architecture is an abstraction of the system architectures based on platforms, programs, protocols, and standards. The reference architecture is defined in domain terms found in the domain model. In fact, the reference architecture can be perceived as an activity-based view of the domain model. An example of a reference architecture is shown in Fig. 2.15.

Using Reference Architectures. The reference architecture is primarily used to instantiate versions of systems, and requires conceptual compliance. For example, assume that two health clinics have different requirements for the size

Triage Task

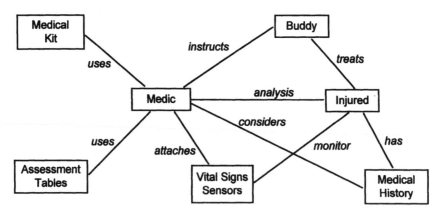

FIG. 2.13. Object model view of a slice (i.e., the triage task) of a domain model.

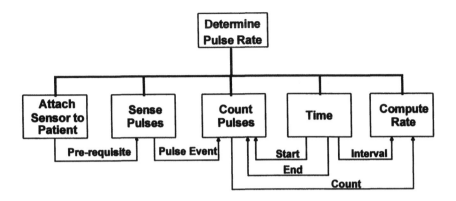

FIG. 2.14. Process model view of a slice (i.e., determine pulse rate subtask) of a domain model slice.

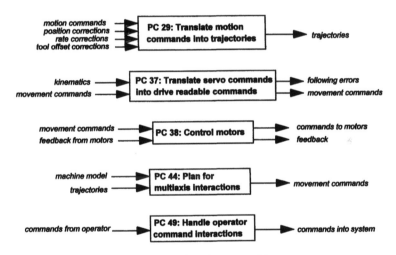

FIG. 2.15. Example reference architecture.

of the databases they use for scheduling patients. The *use* of those databases for scheduling patients is the same. Thus, at the *conceptual* level, the patient scheduling task is the same. Variance occurs only at the implementation level (i.e., in the implementation of two different database components).

One use of a reference architecture is a representation of a conceptual activity representation relevant to domain personnel. Another use is as a tool to obtain compliance. The typical component-based architecture representation has components and interface definitions (as shown in chapter 1). This representation sets the components and their interactions in concrete. For a particular application, this may be the preferred representation.

For a more abstract compliance requirement, argument may ensue on what the components really are, which components communicate with which other components, and what protocols apply.

Our approach to eliminating these arguments is the reference architecture. The primary elements of the reference architecture are *responsibilities*. An element in the reference architecture consists of a responsibility definition, the resources required to carry out the responsibility, the products the responsibility outputs, and the constraints on the responsibility actions. The reference architecture is a set of all these responsibilities. Figure 2.16 illustrates an example.

If there is argument as to the scope of the responsibility, then the responsibility is decomposed into two responsibilities. When the decomposition reverts to computational or performance items, then the responsibility is too detailed and is no longer considered conceptual. This level of detail should *not* become a part of the reference architecture.

Representing Reference Architectures. The reference architecture is represented in an architecture definition language. There are a number of these available. The same language used for the domain model is a possibility. The graphical representation is a set of templates. (The format shown in Fig. 2.16 could be used as a template.)

The reference architecture is used to instantiate applications by applying site or application requirements over the reference architecture responsibilities. The responsibilities may be realized by selecting components based on criteria such as availability, performance, and price. When the components are selected, the interfaces become "hardened" or defined across components.

Reference Requirements

Reference requirements are the conceptual representation or "default template" of requirements. Basic requirements, as revealed during knowledge acquisition, elicitation, and analysis, are depicted. However, the performance values of a requirement field are blank. For example, the text

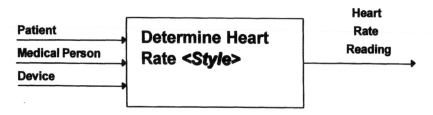

FIG. 2.16. Example responsibility definition.

that follows is extracted from the reference requirements for a machine controller domain. Note that the fact that the system must have a controller, and that the controller has actuators and interprets values from sensors, but may or may not have coordinated axes, is known. The missing values may be different for each version of the system to be developed, based on the needs of a particular site.

Controller

1. The controller *will/will not* have "coordinated axes."
2. The controller shall have # actuators.
3. The controller shall interpret input from #/*type of* sensors.

Using Reference Requirements. Reference requirements are used to define the items of interest in a domain and the possible performance constraints on the domain. They are useful in systems development when multiple versions of systems will be created from one reference architecture. Each multiple system version must meet the requirements of a *specific* site or application. Thus, its requirements will be based on the reference requirements, but will include details on the performance and features specific for that site. The different performance and features are documented as differing values in the fields of reference requirements. Thus, the requirements must be parameterized at the reference level. As the parameter values are selected during requirements analysis for the development of a specific application, the requirements become specific to the application under consideration.

Representing Reference Requirements. Reference requirements may be represented as a template database. These can be linked to the scenarios, tasks, and reference architecture. More formal representations include formal requirements languages used in conventional engineering. The primary guideline is that the representation language selected must support parameterized requirements.

Site or Application Activity Artifacts

The site or application artifacts represent actual or projected situations. Examples are real healthcare clinics, actual machine shops, current classrooms, envisioned banking transactions, visionary electronic commerce systems, and simulated battles.

Site artifacts are more specific and singular in nature than reference artifacts. Commonalties and differences will exist in the information in artifacts for two different sites. The differences across sites represent salient features that must be retained and addressed as the site's information is composed into the reference artifacts. (However, the site artifacts are linked to the derived reference artifacts to create an audit trail.)

Developers review reference requirements in light of site-specific characteristics. They conduct task analysis on significant scenarios to define specific task constraints that apply to a specific site. Using existing domain models and objects identified during task analysis, they conduct object-oriented analysis and performance analysis to develop site-specific models (i.e., specialized domain models and performance models). Using these analyses and the resulting artifacts, developers specialize the requirements, models, and architecture for a particular application. Figure 2.17 illustrates the site specialization process.

Site Model

The site model represents an operational situation or an envisioned situation. It is specific to an environment, a context, and a set of perspectives that must be defined and linked to the site model during documentation. Sequenced processes, specific relationships, and entities with specific, bounded responsibilities can be documented in site models.

While a domain model may contain all the possible relationships or roles for an entity, the site model will represent *only those relevant to the site*.[5] While the domain model of a task may contain a set of subtasks or process steps, the site model will document only the specific process steps actually used and the order in which they occur.

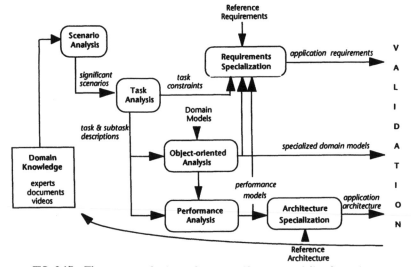

FIG. 2.17. The process of using reference artifacts to specialize for a site.

---

[5]A domain model is comprised of information and objects represented in multiple site models.

Using Site Models. Site models are "where the rubber meets the road," or vision meets reality. Site models are used to help document the model on which the new application will be developed, and to enable communication and refinement. The shareholders at a site can recognize and understand the information in the site model, and can provide feedback and validation to developers. Additionally, the site models are used to synthesize and refine the existing domain model.

Representing Site Models. Site models are represented in forms similar to the domain model and other object models. In some cases, the formal representation languages may differ from the domain model language, however, requiring translators or meta model transitions. Figure 2.18 illustrates the structure of a site model object diagram. The domain model object diagram would be comprised of a number of site model object diagrams.

Site Architecture

The site architecture represents an operational system or an envisioned system that is specific to an environment, a context, and a set of perspectives. During the documentation of activities, developers must define and link the environment, context, or perspectives to the site architecture. For example, the site architecture is comprised of information specific to a

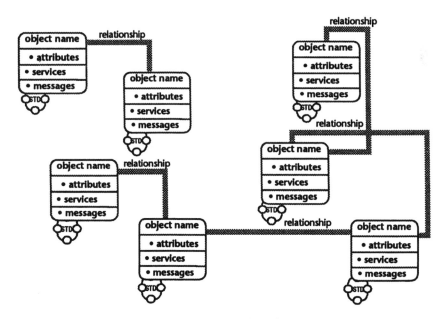

FIG. 2.18. Site model object diagram.

site, including implementation profiles that designate standards selected, interface definitions, and so on.

Using Site Architectures.   The site architecture is used by system developers and component developers to define or select specific components, establish performance values, connect interfaces, and test the requirements. The site architecture is an implementation profile for a specific instance or application of the system. For example, the implementation profile describes choices for the following:

- operating system
- platforms
- communication protocol
- network
- processors
- database, etc.

This profile is valuable information to component developers and users who must specify real or projected systems.

Representing Site Architectures.   The conventional graphical representation of a site architecture is a "boxology" of components and interfaces. A number of architecture definition languages can be used by developers to document the site architecture. To document the application or site architecture, developers work from responsibilities of entities at the site. In fact, one known set of responsibilities may be documented in different ways in the application architecture, depending on the requirements for a particular site. For example, Fig. 2.19 shows two versions of an application architecture. In the version on the left, all responsibilities are executed by a medical person. In the other version, the medical person attaches the sensor, but the automated system executes the remaining responsibilities.

Site Requirements

When engineers of conventional systems conduct requirements analysis and document the results, they are actually documenting the *site* require-

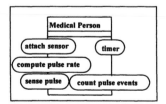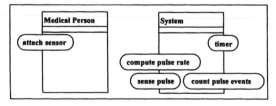

FIG. 2.19. Two application architecture versions.

ments, or those specific to a particular application. In the SEP, site requirements describe the specific site or application requirements requested by the shareholders. Whereas reference requirements are general, site or application requirements are specific. An example of site or application requirements, illustrating how site/application requirements are derived from the reference requirements presented in the previous section, follows:

Controller for [Specific Site/Application]

1. The controller will have "coordinated axes"
   1.1 The controller will have 2 groups of 4 coordinated axes
2. The controller shall have 3 actuators
3. The controller shall interpret input from vision sensors

Using Site Requirements. In conventional systems, development of site requirements involves analyzing and documenting user or shareholder requirements. Thus, each application project "starts from scratch" in the identification and documentation of requirements for a specific application. In the SEP, site requirements are acquired by selecting parameter values for existing *reference* requirements. To develop site requirements, we select the appropriate responsibilities from the reference architecture, based on the elicitation of requirements from shareholders at that site, and the development of a partial site model. These responsibilities are then related to real, specific components. Developers use reference artifacts and site requirements to produce site artifacts, as shown in Fig. 2.20.

Representing Site Requirements. Site requirements are normally represented in textual documents. However, potentially more valuable representations include animations, simulations, user screens, and prototypes.

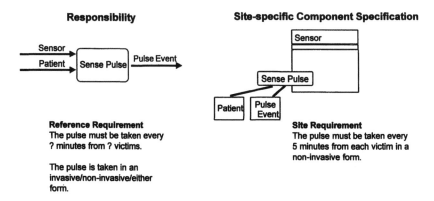

FIG. 2.20. Producing site artifacts from reference artifacts.

Technology or Component Activity Artifacts

All the scenarios, tasks, and behaviors acquired and represented must be carried out eventually by some component. The component may be a human, or it may be automated. If automated, the component may be composed of executable computer code, or it may be specialized equipment, a simulation, or be realized in some other form. In any case, the technology employed is *based* on the situation and is not *forced* on the situation.

Component Models

The component model contains information specific to the domain, which has been acquired and represented by analysts and engineers. This model specifies the coverage, content, and responsibilities of a component. For example, if it is a decision support component for assessing the criticality of wounds, then the component model may contain probabilities of various wounds, decision paths for assessing wounds, and categories of wounds.

Using Component Models. The component model is used to populate the data in the component specification and to define the algorithms and functions for the component. For example, the component model may be used to write the methods for an object in an object-oriented paradigm.

Representing Component Models. Component models are represented in a similar manner to the domain model. In fact, component model information is dispersed throughout the domain model. The following types of information are usually found in component models:

- Attribute information about an entity from a task.
- The methods of entity behavior.
- The state transitions of an entity.
- The communications among entities.

Component Architectures and Specifications

Component architectures and specifications describe the internal structure and action of each component.

Using Component Architectures and Specifications. The component architecture and specifications are used to understand and describe the behavior and algorithms of the component.

Representing Component Architectures and Specifications. Component architectures take the form of specifications of the components. The specifications describe the entities, behaviors, and constraints on the components. Modified flow charts, object models, and formal specifications are representations used for component architectures.

Component Requirements

Component requirements are specific requirements that describe the need for, and functionality of the component. In other words, they describe what a component can do and under what circumstances it operates.

Using Component Requirements. Component requirements are used to match responsibilities and site requirements. Fig. 2.21 illustrates how responsibilities and specifications become mapped to actual components.

Representing Component Requirements. Requirements analysis techniques are used to acquire and describe the capabilities of the component. The same representations used for site requirements shown in Fig. 2.21 are used for component requirements.

Responsibility

Site-specific Component Specification

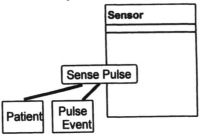

Component

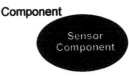

FIG. 2.21. Mapping of responsibilities to component specifications.

SUMMARY

As analysts and engineers begin to explore a domain and conduct knowledge acquisition or elicitation sessions, they produce initial SEP products such as session reports, which document sessions with users, domain experts, or shareholders. Products also may include videotapes and audiotapes of sessions, or domain documents, that are analyzed later. Initial products are used to enhance communication among team members and help developers plan and scope subsequent acquisition and analysis activities in the domain.

Subsequent knowledge acquisition and elicitation activities produce intermediate SEP products including significant scenarios, refined work processes, and task descriptions. Significant scenarios help define the required or desired performance and functionality of a system. Refined work processes are decomposed into the tasks that comprise them. Tasks are reviewed, and priority tasks are selected for further analysis. This analysis produces task descriptions that represent critical information on the priority tasks. Both the scenarios and task descriptions provide crucial input to the primary SEP activities areas—Reference/Conceptual, Site/Application, and Technology/Component.

As engineers and analysts use knowledge acquisition, elicitation, and analysis techniques within SEP activity areas, a variety of products is produced. Primary products produced in the Reference/Conceptual activity area include the domain model, the reference architecture, and the reference requirements. Primary products produced in the Site/Application activity area include models, site architecture, site requirements, and performance requirements. Primary products produced in the Technology/Component activity area include component specifications, models, technology requirements, and performance requirements.

The intermediate products, along with the Reference, Site, and Technology artifacts, comprise the major items produced during SEP activities. The level of effort expended on these items should reflect the extent to which they will be used. Versions of these items should be expected to evolve over the timeframe of the project at varying rates.

Part II

PROCESS & TECHNIQUES

Chapter 3

Planning and Managing Effective Requirements Activities

The cost and complexity of systems have increased during the last decade. Consequently, the cost of errors introduced through incomplete or inaccurate requirements also has increased. Many computer professionals were trained in an era when we defined requirements by identifying and defining data used in transactions. However, today we also must attend to how performers use knowledge and higher level thinking skills to act on and manipulate data and information. Today's complex information systems often are required to have both traditional and knowledge-based components. Additionally, we may be required to design them so architectures, components, and devices are reusable. In this systems development environment, the task of planning and conducting requirements elicitation and analysis is more critical than ever.

The purpose of this chapter is to assist project personnel in planning and managing requirements-gathering activities. This chapter is intended to help readers achieve the following:

- Avoid common problems associated with requirements gathering, knowledge acquisition, and analysis.
- Manage risks by putting appropriate plans in place early in a project's life cycle.
- Organize and prepare the project participants and tools.
- Manage the process and logistics, including planning and organizing onsite sessions, conducting sessions, documenting sessions, and ensuring adequate review of session information.

AVOIDING COMMON PROBLEMS
AND MANAGING RISKS

To effectively conduct successful requirements-gathering and domain analysis activities, we must understand, anticipate, and plan to avoid common trouble spots. Some of the more commonly discussed problems include:

- Failure to ensure that systems are user-centered.
- Performers who provide marginal quality of expertise or whose responses are inaccurate.
- Inadequate access to key performers at appropriate points in the project.
- Inadequate time scheduled for requirements activities.
- Ineffective communication with performers.
- Uncooperative attitudes or resistance on the part of performers.
- Failure to document requirements.

The following sections expand on the impact of each of these issues and provide approaches to mitigate risks, and a foundation for the procedures and suggestions presented in subsequent sections of this chapter.

Failure to Ensure That Systems Are User-Centered

Failing to ensure that systems are user-centered may result in unhappy users or incomplete projects. When information technology managers were asked why projects failed or were canceled, they listed lack of user involvement and incomplete requirements as the top reasons (Johnson, 1995).

We believe these factors are related, and that appropriate user involvement will enable developers to define more accurate, complete requirements. How does a project ensure that a user-centered system is developed? As described in chapter 1, the Scenario-based Engineering Process is user-centered because it relies on the solicitation and analysis of scenarios from within the work context. Additionally, the systematic requirements elicitation and analysis activities we propose in later sections of this chapter depend on the involvement of performers throughout the process.

Marginal Quality of Expertise and Misinformation

For requirements to meet performers' needs, analysts must be able to elicit necessary information and heuristics. Poorly planned requirements elicitation may result in inaccurate requirements—"garbage in, garbage out." Managing this problem requires attention to several issues. First, we must ensure that we identify the right performers with whom to work. To help ensure good information quality, customers should select expert performers to help guide the process and validate the requirements. These performers can work with analysts in sessions, but are even more useful as members of a task force that helps to plan sessions, select performers for sessions, and evaluate information gathered.

Second, we must recognize the prevalent knowledge types in the domain because some types of knowledge are more difficult to extract than others (see chapter 4). Asking a performer to explain how he or she makes a decision invites misinformation because people are not normally cognizant of deep reasoning processes. The accuracy and quality of the information we elicit is likely to be better if we present the performer with a problem to solve, and track decisions and factors considered within the context of the problem.

Third, asking performers to review session reports in which they participated, to edit them, and to validate their accuracy also positively impacts quality. This gives performers an opportunity to correct mistakes analysts made, to correct misinformation they may have inadvertently provided, or to provide information they recalled after the session's conclusion.

Finally, use a task force which consists of the project team and several experts to review and validate requirements as they are gathered. (The task force is described in subsequent sections of this chapter.)

Inadequate Access to Key Performers

Scenario-based engineering requires user-centered design, and users come in all shapes, sizes, and levels of expertise. Projects must ensure that they tap a diverse group of users—experts, novices, and marginal performers— to enable the system to support all user types. However, to specify and design screens and functionality that enhance the work process, project teams must focus on the people whose performance the customer most wants the system to emulate. If the project team can't get access to the performers who are more likely to provide high quality, complete, requirements-related information, user-centered design is almost impossible.

Access can be a problem on several fronts. First, top performers—the experts with whom we want to work—may be available, but may also be geographically dispersed across the country or the world. Gaining adequate access to these performers requires careful planning. One solution may be to plan a travel budget that supports requirements sessions in representative sites, to enable analysts to gain access to specific performers. Once on site, analysts use carefully selected techniques, and videotape interactions with key performers to make the most of the session opportunities, and to enable later analysis. If travel budgets are tight we can use videoconferencing and, to a lesser extent, teleconferencing. We have successfully used videoconferencing on a number of commercial and government projects and found that once analysts and performers learn how to communicate with each other through this medium, it is quite effective. In fact, unlike physical site visits, which may require the expert to be available to the analysts all day, videoconferencing enables analysts to meet with experts in a number of shorter, more manageable sessions requiring smaller segments of the expert's time.

Access problems also occur when experts are unavailable, for example, if the customer is reluctant to take expert performers "off the job" to spend time in sessions. Consequently, analysts may be sent to work with an alternate performer—one who is more accessible but who does not embody the expertise and experience the project needs. We believe the best way to avoid this problem is to document the requirement that the project have adequate access to performers (expert and novices) in the initial proposal. During the project alignment meeting, reiterate the importance of access to appropriate performers and define your expectations for their time. Presenting the client with a schedule indicating approximate dates for sessions and an estimation of performer time per session often helps avoid access difficulties later.

Inadequate Time Scheduled for Requirements Activities

Recent research estimates that only 16.2% of American companies and government agency software projects were completed on time and on budget in 1995 (Johnson, 1995). The more complex the project, the more at risk it is for schedule and budget overruns. While most projects schedule enough time for development, integration, and testing, there is a tendency to plan less time than is actually needed to elicit requirements and define specifications. This may be the case because in the past, many contractors were often given requirements by the customer/client. However, in today's environment it is probable that the customer doesn't really know what is needed at the level required for development. Consequently, not only must we define adequate planning and requirements phases, but we must also plan for iterative refinement on initial requirements. With the advent of new development trends that include the creation of proofs-of-concepts and prototypes (both visual and functional), the line between requirements and design blurs. This results in a longer requirements phase that overlaps a portion of the design phase.

Yet in planning requirements elicitation activities, many projects schedule time only for minimal site visits or interaction with a very few performers. We suggest planning to gather requirements from numerous sites, working with a range of performers, and using a number of different techniques to ensure they can create adequate, complete requirements. This schedule should include time for:

- Upfront domain familiarization.
- Initial site observation and requirements elicitation sessions.
- Analysis and production of session output.
- Subsequent elicitation sessions to firm up initial requirements or elicit information for previously missing requirements.
- Development of a proof-of-concept, sample screens, or a visual prototype and presentation of the initial visualization of requirements to elicit feedback from performers and to extend requirements.
- Documentation of requirements.

Ineffective Communication With Performers

Once analysts do gain access to and begin working with performers, the way they plan, conduct, and review sessions must be professional. Communication is the key to making this happen. Ineffective communication occurs when analysts

- Fail to learn basic domain vocabulary and acronyms.
- Don't provide performers with adequate explanation of why they want to conduct sessions with them.
- Fail to present sessions and the project in a way that ensure performers there's "something in it for them."
- Ignore session introductions and begin sessions without giving the performer a context.
- End a session without summarizing what was accomplished, a general understanding of information gathered, and an idea of what happens next.
- Take over a session and do more talking than the performer.
- Fail to use active listening or verbal and nonverbal feedback.

The result is that performers can't or aren't willing to give the information needed because they sense analysts don't care, don't understand, or don't have their best interests at heart. To avoid these situations, analysts must be good, mature communicators who have demonstrated their listening skills, questioning skills, and feedback mechanisms.

Uncooperative Attitudes or Resistance

Rarely do we encounter uncooperative attitudes or resistance from performers who have been well prepared. Under normal circumstances, once performers understand the approach they are eager to participate and share their ideas. We believe the SEP requirements elicitation and design actually diminishes resistance because performers feel they are part of the process. More and more often, computer systems are a critical part of their jobs and if designed well, these systems can enhance their opportunity for success. Consequently, they want to have a say in new functionality, screen redesign, work flow, and so on.

However, on occasion an individual performer may appear uncooperative or resistant. When that happens, examine the situation and look for motivation to cooperate. Although the project may have been granted access to performers, they may still have to produce the same amount that day as any other work day. Working with analysts may increase their workload while decreasing the amount of time in which they must complete their tasks. To counter this factor, address it in the alignment meeting. Can quotas be lessened on the day or week when performers assist in sessions or reviews? Can they log into the system under a charge number that shows them off-line and doesn't figure into their performance measures?

Performers' reluctance may also reveal insecurity in themselves or a lack of faith in the process. They may fear they don't really know all that is needed, or that they can't possibly convey what they know to an outsider in such a short amount of time. To address these issues, analysts should describe their role in the total process, noting that other analysts are meeting with performers at other sites. Also, let performers know it is acceptable to say "I don't know" or "I'll try to find out" during the session.

Finally, some performers may be resistant or uncooperative because they fear job loss or less job security. In today's environment, these are realistic fears that must be addressed. As an outsider who is working with a client/customer group, an analyst cannot assure them safety. However, we have found that a memo from upper management describing the system development project, its goals, and future plans can make a big difference. If, because of past or expected downsizing this problem is severe, we may suggest that the client combine the system development project with a "change management" project. This can help performer groups define and embrace new business processes, procedures, and system plans.

Failure to Document Requirements

This problem area assumes that some type of formal deliverable will be prepared. However, a formal requirements document isn't enough to eliminate the risk involved when multiple analyst teams elicit requirements from individual performers and groups of performers at more than one site. The failure to which we refer relates to initial and intermediate documentation of requirements. Imagine that you send three teams of two professionals out to conduct sessions at three different customer sites—call or service centers, hospitals, stores, etc. Together, they meet with 24 individual performers, ranging from novices to experts. Additionally, they conduct focus groups to validate initial functional requirements and a new work process. Now imagine they returned to your site but failed to summarize their sessions in detail because the schedule didn't allow it. Next week, two of them are placed on a "tiger team" to save an at-risk project. Or worse yet, one of them resigns or becomes disabled. At this point the project has lost valuable firsthand information from performers. The approach detailed in this text, and the procedures outlined in this chapter specifically, suggest that all sessions be documented in a way that enables them to be preserved, reviewed and analyzed by other team members, and communicated to clients.

The sections that follow present procedures and plans to help avoid the aforementioned problems, including organizing and preparing project participants and resources, and managing the logistics and process.

ORGANIZING AND PREPARING PROJECT PARTICIPANTS AND RESOURCES

Recently a group of experienced knowledge engineers and analysts were discussing common difficulties related to requirements gathering and analysis. Each of them agreed that once you learn the techniques and feel confident using them the major problem is simply the logistics involved in preparing, planning, and organizing requirements-gathering activities. The suggestions that follow can help large, complex projects avoid or minimize common difficulties.

Conduct Project Alignment and Ensure Performer Participation

Project alignment is a management function that impacts the eventual success of the requirements elicitation efforts. The client organization and the group that will be building the system should agree on the purpose, goals, major activities, deliverables, schedule, and acceptance criteria prior to beginning project activities. Figure 3.1 illustrates the alignment meeting and a sample agenda.

The project alignment meeting can be an effective forum for ensuring that clients understand the Scenario-based Engineering Process and the requirements elicitation and analysis activities, specifically regarding the importance of involving client performers in these activities. During project alignment activities we suggest discussing and seeking agreement on the following requirements-related issues:

- Expected type of site visits or sessions required (i.e., interviews, interactive observations, etc.) and the overall purpose and goals of these sessions.

FIG. 3.1. Project alignment helps ensure requirements elicitation success.

- Type of performers (i.e., supervisors vs. performers, experts vs. novices) who should be involved.
- Estimated coverage to ensure sessions cover appropriate sites (i.e., regional differences in sites may require a set of sessions at sites in each of the four regions).
- Identification of a client liaison to coordinate interaction between the analyst team and individual performers (e.g., scheduling sessions, explaining the process, securing domain materials, etc.).
- Estimated schedule for conducting sessions to help client liaisons plan and manage the logistics of scheduling analysts into various sites.

Develop a Plan to Gather, Analyze, and Document Requirements

A plan helps answer logistics questions such as "Who does what, when?" A large team comprised of multiple developers, analysts, domain specialists, and managers may all have different expectations for requirements elicitation and interpretations of the techniques that will be used. A project plan should include sections that detail the following:

1. *Identification of subteams and their responsibility.* On large projects it may be difficult for all analysts to conduct sessions in every site required. The

system may be too complex, it may be physically impossible to travel to each site, or the schedule may be constrained. Under these circumstances, project management may wish to team analysts and developers to specialize in sessions within a particular region of the country, or on a particular functional area of the system.

2. *How requirements will be elicited and gathered.* Identify the approach to be used to elicit and gather requirements. This should include a discussion of the depth and level of specificity, specific goals, and how the information will be used. Additionally, it should document the team organization. For example, this information should detail logistics such as:

- Have domain specialists been hired to support the project?
- Are analysts being paired with domain specialists, developers, or other analysts?
- Will analysts be required to work in teams with client personnel and if so, what is their responsibility (i.e., scribes, analyst support, learning the process)?

3. *Techniques that will be used throughout the project.* Of the total set of elicitation and analysis techniques, which are acceptable for use on this particular project? How is each selected technique defined? Under what circumstances should a particular technique be selected? What difficulties are associated with a technique, and how can the analyst mitigate risks involved it its use?

4. *How techniques will be used and documented.* For each type of technique that will be used, what requirements exist? In other words, what is required of the analyst who conducts each type of technique? What procedures should be followed to ensure effective sessions using a specific technique? What type of documentation is expected of each technique—formal, complete transcriptions, or summaries of session activity?

5. *Desired output per session type.* For each type of technique that will be used, what kind of output is expected? (See chapter 4.) For example, if a concept analysis session is conducted, in what format should the concepts and concept hierarchies be presented? If a task analysis session is conducted, is a task/subtask matrix always required, and what should it include?

6. *Using tapes to preserve session content.* Is it desirable to tape (e.g., audio or video) elicitation sessions on this project? If so, what procedures and guidelines are in place to ensure performer confidentiality of information? How will the team ensure it has the performer's agreement, enabling analysts and developers to share session tapes? Who will store the tapes, how should they be labeled, and how can they be checked out for review?

7. *Procedures for setting up sessions and conducting and reviewing sessions.* Who is the client contact? Who arranges sessions—an internal coordinator,

the client liaison, or individual analysts? Who must approve sessions? What type of session plan or description should be sent to performers to elicit their involvement in the session? What guidelines should be followed to conduct each type of session? (See chapter 4.) How should session material be reviewed, verified, and validated prior to being accepted and used for system requirements and design?

Develop Standard Forms, Templates, and Document Storage Mechanisms

How will analysts keep track of session goals, performer participation, and session results? Once a session has been conducted, in what format will analysts document it, and how and where will it be stored for access by others on the program? To ensure that team members can share the information collected during sessions, we recommend that each project define standard templates for session report forms and technique output. As a new session is planned, the analyst imports the session report template and fills in basic information about the session. The top part of the forms we use includes the following basic information: names of the analyst and performer, the time and date of the session, the location of the session, the type of session, and the session topic and goal. Figure 3.2 provides a sample session report form template. The type of information included in the body of the form will vary per session and session goals, but should include at least organized session content, open issues, work artifacts or forms collected, and follow-up tasks (see chapter 4). After the session has been conducted, the analyst uses this form to translate and document the session content in an organized fashion.

For most of our projects we have stored these report forms in a standard word processing format such as Microsoft Word.® Consequently, reports can be read and used across platforms, imported easily into document management servers or databases, or converted to Web-accessible formats.

As the analyst prepares a new session report form, he or she uses the template and saves the new session with a unique identifier. This identifier is used to track the session, store the completed session form in a database or on a server, and to locate the session for later access and retrieval. (For example, if nine sessions are conducted at a Baltimore call center, they might be saved as B-101 through B-109.)

Standard Output Formats

If a team is large and/or geographically dispersed, consistency and coordination among the team members is critical. It is imperative that regardless of which analyst conducted a session and its analysis, the resulting output is in a standard format that can be shared and understood by

```
┌─────────────────────────────────────────────────────────────────┐
│                         Session Report                          │
│                                                                 │
│  Session #:              Session Date:       Session Time:      │
│  Knowledge Engineer:                         Performer:         │
│                                              Ph:                │
│                                                                 │
│  Session Location:                                              │
│                                                                 │
│  Type of Session:                                               │
│  ___ Interview          ___Task analysis      ___Scenario analysis │
│  ___ Concept analysis   ___ Observation       ___ Structured interview │
│  ___ Work process analysis ___ Process tracing ___ Other:       │
├─────────────────────────────────────────────────────────────────┤
│ General Topic Area:                                             │
│                                                                 │
│ Session Objectives/goals:                                       │
│ 1.                                                              │
│ 2.                                                              │
│ 3.                                                              │
├─────────────────────────────────────────────────────────────────┤
│                        Session Summary                          │
│ Topic:                                                          │
│ Subtopic 1                                                      │
│ Subtopic 2                                                      │
│                                                                 │
│ Topic:                                                          │
│ Subtopic 1                                                      │
│ Subtopic 2                                                      │
│                                                                 │
│ Scenarios identified:                                           │
│                                                                 │
│ Domain materials received:                                      │
│                                                                 │
│ Open issues and followup tasks:                                 │
└─────────────────────────────────────────────────────────────────┘
```

FIG. 3.2.   Sample session report form template.

all team members. As chapter 4 describes, standard output formats depend on the techniques used. Each project should identify the acceptable output formats for each technique that will be used. We suggest doing so early in the project and creating templates and examples of these output formats, which are placed on a file server for easy access and use by all team members.

Session Report Storage

The reports that are produced after each session provide ongoing documentation of interactions with performers, and reveal system requirements and design issues. Session reports constitute critical work artifacts of the elicitation and analysis process and must be available for all team members to help them prepare for upcoming sessions, review work by peers in a particular area, help new analysts "bootstrap" up quickly, or enable review by verification and validation (V&V) teams.

We suggest keeping session reports filed electronically, either in a database, a document management system (i.e., Lotus Notes), or on a document server. For example, one project set up a document server to manage all the session documents created by project personnel, providing the following functions:

- Browse all documents.
- Browse documents by topic.
- Search for a document.
- Add, modify, and delete a document.
- Add, modify, and delete a topic.

Analysts used the Internet and a browser (e.g., Netscape) to access the session reports and the related documents they needed. They could view, download, and print the original document. This arrangement enabled electronic access by all team members, electronic searching, the ability to copy and print selected sessions, and the ability to sort sessions by type, performers, analysts, etc. to assist in report production.

Establish Session Procedures

Determine what procedures analysts should follow in planning sessions, contacting performers, conducting sessions, transcribing and organizing sessions, ensuring sessions are reviewed, and posting sessions to a project-wide server or database. Provide guidance to ensure that analysts use videotapes or audiotapes appropriately and maintain good relations with performers.

Establish a Reference Center

Imagine a large project with multiple analysts and developers, each of whom has been assigned specific tasks or areas of responsibility. Many of them will need information that is unique to their assignment; however, all of them will need general information about the domain and its primary concepts, vocabulary, and organization. Furthermore, as the project matures, personnel changes may necessitate that new people replace original analysts, or that responsibilities be modified. In each case analysts would most certainly duplicate efforts as they research and request domain-related information including training manuals, procedures manuals, organization charts, mission statements, and documentation on legacy systems. Establishing a project-wide reference center can eliminate duplication, reduce costs, and reduce the number of requests to the customer for information that may already be in house. (It's embarrassing to ask a client for a document and be told, "I gave that to Sue in your organization last month.") Once a reference center is in place, it can house materials that can be used to develop document frameworks, prepare for sessions prior to meeting with the customer's performers, and store hard copies of notes or transcripts from sessions.

Establishing a reference center isn't hard, but it does require a commitment by project management for personnel, time, space, and in some cases, acquisition costs. The following steps summarize the major tasks involved.

Determine Who Functions as the "Librarian"

Every reference center requires the part-time services of someone who will ensure that it gets and stays organized. We often appoint one of our more junior analysts to manage the reference center. This involves establishing a working relationship with the client contact most likely to provide the target documentation, following up to secure materials promised but not yet received, and compiling requests from the analyst team so they can be presented in a single memo. It also involves maintaining the files in an organized (i.e., easy to find) fashion and keeping a current file of what is in the reference center.

Pick a Location

Select a central physical location for the reference system. If the project is very large, a small room or cubicle with an area for filing and a desk for reviewing materials may be required. Otherwise, a large file or set of filing drawers may suffice. Overestimate the space you will need—your collection will grow quickly!

Develop the Filing System and Database

Catalog materials as they are placed in the reference center. Decide how you want to categorize the materials—based on work process or function, type of material (training, videotapes, policies/procedures, etc.), and so on. Set up the files accordingly, using tabs that quickly help someone find an area or category. Make browsing easy by either organizing documents within a category alphabetically or by numbering each document. Numbering a document within a category makes it easy to see which documents are missing and to replace documents quickly.

Create a simple database for the project that lists very basic information about each reference item. Suggested fields include: title, author, publication date, key words for searching, a short abstract, and to whom it is currently loaned. Figure 3.3 provides an example of a Filemaker Pro® database that logged over 150 items for a large project. It was kept on a file server to encourage easy access and use by all team members. Its structure made it easy for team members to find the documents that were of interest to them. In addition, this database helped the project track and maintain the materials after they had been loaned to someone. On a quarterly basis, a hard copy of the report showing documents on file can be produced.

Determine a Customer Contact Person

If you are requesting a large number of documents at once, select a contact person on the customer's team who will be responsible for obtaining requested materials. Establish dates for having that material in house.

Determine Procedures for Requesting Materials and Notifying Participants of the Arrival of New Materials

Procedures for requesting materials help ensure that each team member is not inundating different performers and customer contacts with requests for the same item—or worse yet, requesting a document someone else has

Item Number	116	**Publication Type**	Form
Reference Number			
Title	Patient Status Report		
Author	Baltimore County	**Date**	1991
Origin Organization	Baltimore County Fire Department, EMS		
Domain	Urban		
Description	Small form used for recording patient status, medical history, treatment rendered, and medications given.		
Keywords	status, history, EMS		
Date Acquired	11/14/94		
Source			
KA Event	Metro Blitz	**Session**	N/A
Status	in library	**Loaned To**	
Loan Date		**Return Date**	

FIG. 3.3. Sample entry from a reference materials database.

already requested or received. This can be handled in a number of ways, including the suggestions that follow.

- Prior to requesting a document, check the reference database to see if a copy already is on file.
- Send e-mail to your project team identifying the document and requesting team members to respond if they have already requested or received it.
- In weekly reports on progress and plans, include a section that identifies documents or materials needed.
- Team leader or designee compiles requests for documents/materials and submits them to the customer contact.

If a training document, job aid, form, or other work artifact is provided or identified during a session, the analyst is responsible for either delivering the item to the reference center after the session or following up after the session to obtain the item.

Develop "Check Out" and Return Procedures

After a team member finds a needed document, there should be an easy way to track who borrowed the item and when. If the team is located in a single place, this can be easy as a log form in the reference center in which the team member records his or her name, the date, and which document was taken. If the team is geographically dispersed, the team member locates the item and makes a request for it via telephone or e-mail. The person responsible for the reference center verifies that the material is in house and sends it (or a copy) to the requester, along with a note suggesting a return time. If the material is not in by the suggested time, the reference center contact must follow up to encourage its return.

Train Analysts on Techniques, Process, and Domain Basics

Prior to conducting site visits, conduct an informal needs assessment to determine the types of training the analysts on your team require. At a minimum, analysts should be able to:

- Conduct and manage sessions using each of the selected techniques and be able to document session output.
- Analyze the output of selected techniques to produce the desired output.
- Interact effectively with performers (including communication, feedback, etc.).

- Communicate and understand terminology used to describe the domain.

Obviously, analysts must be capable of conducting sessions with each of the techniques selected for use on the project. Project management must feel confident that all analysts have completed training in each technique and can meet an accepted standard in its use. For example, several companies with which we worked asked us to present a 3-day workshop in requirements elicitation and analysis techniques in which all team analysts would participate prior to going onsite. A group workshop like this enables all analysts to learn the same information about each technique, and provides a common standard for understanding and using the technique. Additionally, an interactive workshop gives all analysts the opportunity to review guidelines for techniques and to practice them in pairs or small groups to boost analyst confidence and technique effectiveness.

Although each of us communicates daily, analysts also may benefit from training in communications—specifically listening, facilitating, and using verbal and nonverbal feedback. The ability to communicate effectively during sessions is critical to the success of each technique. Reviewing this material during a training workshop devoted to techniques is extremely beneficial.

Additionally, analysts must be trained to produce the output desired. This output varies according to the techniques selected (see chapter 4 and individual chapters on techniques). Even if analysts have conducted a task analysis before, project management cannot assume they know how to analyze and document session output or produce a task hierarchy diagram consistent with the format used by other team members. Teaching analysts what is expected and providing templates and examples during training increases the likelihood of producing usable output.

Training should also address the need for analysts to have some understanding of the domain's key vocabulary and acronyms. Although analysts need not become experts in the domain, they should learn enough to enable them to ask intelligent questions and understand the responses to those questions. This type of "training" may be accomplished on an individual basis by preparing a reading list that includes important, basic domain material. Each analyst could be expected to complete this list prior to conducting sessions. In other cases it may be useful to provide a group lecture or presentation that walks analysts through a company's organization, mission, critical success factors, competition, and descriptions of work environments or performer types. Prerau (1987) suggested bringing in an expert performer or a representative of the organization who can provide a tutorial presentation and review domain terminology for analysts.

MANAGING THE LOGISTICS AND PROCESS

Many requirements elicitation "horror stories" relate to managing the logistics and process. Some examples include:

- Site managers who didn't know the team was coming and therefore weren't prepared to provide facilities or personnel for sessions.
- Analysts who arrive at a site only to find the key performers involved in an off-site strategy meeting or inservice training session.
- Sessions conducted in only one site or geographic area, resulting in a biased view. Performers and managers of other sites attack the completeness or accuracy of the system requirements.
- Performers and supervisors to whom information about the purpose of the visit and general information about the system under development had not been communicated.
- Union shops that balked at cooperating because their personnel didn't fully understand how the information would (and would not) be used.
- No firm schedule describing which analyst was meeting with which performer, when, or where, resulting in missed meetings, less time spent in sessions, and a perception that the analysts are unprofessional.
- Poorly trained analysts who were unfamiliar with the techniques or tools used in the sessions and who failed to get required information at the appropriate level of detail.
- Techniques that require videotaping although no one requested permission to tape beforehand, resulting in an inability to tape or reluctance on the part of performers sensitive about taping.
- Large, multianalyst requirements-gathering and analysis activities without the use of standard tools (i.e., forms or expected output formats). This results in information that is documented in various ways and at different levels of detail, and communication difficulties among the team.
- Projects on which the analyst conducted sessions as planned, but failed to document information gathered in a way that enabled easy access and comprehension. The analyst later left the company and sessions had to be conducted again to fill in the resulting information gaps.

To avoid these problems, carefully select performers for the task force and sessions, plan and schedule sessions, and conduct sessions professionally. Then document sessions in a standard fashion, maintain a domain dictionary, and adhere to review procedures.

Set Up a Task Force and Select Performers for Sessions

We suggest establishing two different groups of performers with whom sessions are planned and conducted. The first group consists of a task or review board comprised of representatives from the development team and the customer's organization. Whereas some of these individuals may be project management contacts, others should include carefully selected expert performers and supervisors. These performers should represent the types of professionals who will work with, or be impacted by, the system under development. For example, if the new system will support two types of telephone service agents—those responsible for international calls and those responsible for domestic calls—at least one international agent and one domestic agent should be members of the task force. A supervisor who can represent management's perspective could be chosen from either area.

All members of the task force should receive an orientation that includes an introduction to requirements elicitation and analysis within the Scenario-based Engineering Process, with an emphasis on the importance of scenarios and user-centered design. This task force should meet on a regular (i.e., monthly) basis to accomplish the following:

- Establish goals for requirements elicitation activities (e.g., what must be revealed, fine tuning the level of specificity).
- Review and refine plans for requirements elicitation.
- Review progress against goals.
- Assist in selecting sites (e.g., determining factors important in site location, creating a potential site list, rating and ranking sites, obtaining cooperation from sites).
- Assist in identifying and notifying performers for sessions.

In addition, members of the task force are instrumental in serving as liasons with the field to help plan and structure onsite activities, and working with site representatives to ensure session logistics are well managed (McGraw, 1994d).

Other performers who will be identified will comprise the "pool" from which analysts and/or the requirements coordinator will select people to participate in specific sessions. Task force members help identify performers with whom sessions can be conducted, or identify supervisors who could submit names of appropriate performers. As they are identified, performers' names and basic information should be entered into a database that will enable the analyst team to sort based on name, expertise, site, and so on. When the need for a session is identified, the coordinator can assist in selecting a performer who might have the expertise needed. Figure 3.4 is an example of a database used for this purpose.

Margaret	Rutherford	R.N., C.E.N.
DE First	**DE Last**	**DE Title**

The Washington Hospital Center	(202) 345-9867
Organization Name	**Phone/Extension**
Critical Care Services	(202) 345-2461
Department	**Other Phone**
110 Michelson Ave., NW.	
Address	**Fax Number**

Washington	DC	20010	Dr. Ralph Christensen
City	**ST**	**Zip**	**Referred By**

Expertise Flight nurse

MCC Approval ⦿ Approved ○ Pending ○ Not submitted

Training Coordinator for medevac

Notes

| Letter | Envelope | Fax | CallLog | Add Name | Find | List | Labels |

FIG. 3.4. Sample database for managing performer contacts and selecting performers for sessions.

Plan and Schedule Sessions

The task force assists analysts in planning the initial number and type of sessions required—interviews, process tracing, observation, etc.—and the general schedule for session activity. After potential performers or contacts have been identified the analyst begins to plan and schedule individual sessions. Planning a session involves the following:

- Identifying session goals and topics and preparing required forms or worksheets.
- Working with the client contact or the project's coordinator to identify an appropriate performer.
- Contacting them to establish an appropriate date and time.
- Helping the performer prepare for the session.

First, the analyst reviews the project schedule and goals to define the type of information required, the desired level of detail, and the session objectives. For each session, the analyst should complete the top part of a session report form to document these goals and desired results. Figure 3.5 is an example showing the top part of a completed sample session report form.

Next, the analyst works with the contact person (the project coordinator, liason, or other client contact) to identify an appropriate performer with whom to conduct the session. On large projects we keep a database of available performers, which enables us to sort by expertise, geographic area, and so on.

After the analyst has identified a candidate, he or she contacts the candidate and/or his or her supervisor to discuss availability. If the analyst is working with a client contact to select performers, we suggest following up with a memo to help the client contact select performers and elicit their participation. The memo should address the following questions:

- *What is the purpose of the session and what is the long-term goal (i.e., why is our organization developing this system)?*
  A purpose statement helps answer the "what's in it for me" question and helps motivate performers to participate and provide input. A simple description of the system development activity can be included in the memo. The memo should also state that the analyst wishes to get input from them to help design a system that will make it easier for them to do their jobs, support them in performing work, etc. Help

---

Session Report

Session #: M-133   Session Date: 3/23/95   Session Time: 1:00-3:30
Knowledge Engineer: K. McGraw   Performer: J. Smith
                                                       Ph: (410) 756-3948

Session Location: EMS Headquarters Bldg., Mitchell, MD

Type of Session:
 _X_ Interview   ___Task analysis   ___Scenario analysis
 ___ Concept analysis   ___ Observation   ___ Structured interview
 ___ Work process analysis   ___ Process tracing   ___ Other:

General Topic Area: Scene Assessment

Session Objectives/goals:
1. Identify tasks and decisions performed during scene assessment
2. Identify information used to assess the scene
3. Identify scenarios that illustrate scene assessment

FIG. 3.5. Sample planning and identification section of a session report form.

them understand the notion of user-centered design. Once they understand how their input will be used, most performers are glad to assist.

- *Why do they want to talk to or work with me, and what will the experience be like?* Performers are often intimidated by words like "task analysis," "user-centered design," "process tracing," and "concept analysis" because they may assume it means they are being evaluated. Others may be excited initially to be included, but later may fear they won't know what to say or do. Describe briefly in the memo what will occur in the session and the kind of information sought, to help alleviate some of these initial concerns. Give them examples! If the session is a work process analysis, for example, send out a work process matrix along with the memo and ask them to fill it out or think about it prior to the meeting. This will indicate the format and depth desired.

- *Where and when should these sessions occur?* Describe where the sessions will occur—in the context of their workplace or their office, or in a conference room. Let them know when the sessions will begin and the approximate length of each session. This will enable them to plan around the sessions.

- *What do the analysts expect of the performer?* Performers should be told to be open and forthright, to act naturally, and to treat the analyst as an apprentice who will ask questions and look for detail. If analysts expect any additional follow-up activity (i.e., reviewing session notes, etc.), this should be stated.

- *How can performers prepare for the sessions?* Different sessions can benefit from different types of preparation. Tell the performers what they can do to be most effective. For example, suggest that performers think about interesting or challenging cases. If they have examples, forms, reference material, or other material that would assist them in describing an activity or scenario, they should be encouraged to bring them.

- *Who can they contact if they have more questions?* During the project alignment meeting or a task force meeting, appoint someone who can serve as the primary liaison and communicate with performers. Include that person's name and number in the memo, or ask them to send the memo. If performers are nervous about upcoming sessions, or unsure how to prepare, they know who to contact.

We have communicated successfully with performers to plan sessions by telephone, e-mail, and fax. Figure 3.6 is a sample memo to help the domain contact select the right people for a session, and help the people prepare for the session in advance.

To: Lt. Smith, Fire/EMS
Re: Upcoming sessions at your facility

Thank you for agreeing to let me conduct 3 requirements gathering sessions at your 401 Beakman Ave. site this Friday, starting at 3:00 PM.

The purpose of the visit is to conduct further analysis that would help us accomplish the following goals:
- Understand the format and type of information you receive from 911 calls
- Document how you interact with your computer system to log those calls
- Break down the tasks involved in communicating that information to units
- Analyze the decision making involved in selecting the appropriate response units
- Monitor your interactions to determine the type of information you receive back from the units

I would like to spend approximately 1 hour with each of 3 performers who fill the following types of positions:
- Performer who gets the 911 call
- Performer who handles fire and emergency calls and selects equipment and unit(s) that will respond
- Performer who communicates with and monitors units as they respond

If time allows, I will be very interested in seeing anything else you may believe would help us understand this process. Thanks in advance for agreeing to support this important project activity.

FIG. 3.6. Sample memo to help a contact select performers and prepare them for the upcoming session.

Figure 3.7 presents an example of a memo that went to individual performers to help them prepare for scheduled sessions.

Conduct Sessions

Manage the Environment. Managing and controlling the environment of a session requires many pre-session activities, such as:

- Confirming the session with the performer.
- Reviewing materials to prepare for the session.
- Reserving a room or obtaining a room assignment (and directions).
- Arranging for appropriate taping equipment or props (e.g., cases, examples, etc.).

Prior to the session start time the analyst should go to the room to ensure it is conducive to the type of session planned; any required equip-

To: Session Participants
Fr: Betsy Crandall (395-8374)
Re: Upcoming sessions for Patient Management System

Karen McGraw and Jill Loukides will conduct an interactive observation session with each of you that will last from an hour to hour and a half. During the session they will be interested in observing and talking to you about:
• Relevant patient information captured and entered and the role you play in this process
• Tasks you have to perform to produce the patient record
• Interaction with others required to accomplish these tasks today
• Examples of scenarios that illustrate the normal process and problems.

Sample questions they might ask include:
• What is your primary job role or responsibility?
• (Looking at sample reports and forms . . .) Is the information categorized efficiently? Is the information presented efficiently for your job function?
• What information is MOST important on the forms you currently use? LEAST important?
• What problems exist that inhibit you from entering patient information easily?

How to prepare for the session
  Think about the issues outlined above. Collect forms you use to record information and prepare to show them what you do to gather patient information. Consider special cases or examples you should share to demonstrate difficulties or problems. Also, think about your wish list, or features that would make the process easier or more efficient.

Schedule
  A schedule of sessions is attached. If you are unable to meet for any reason, or have questions, call Betsy or Karen McGraw directly.

Thank you in advance for your participation!

FIG. 3.7. Sample memo to participants.

ment should be tested. Environmental factors that can influence the effictiveness of a session (and the participant's perception of its professionalism) include:

• Room size and equipment—the room should be able to comfortably accommodate 3 to 4 people. Equipment needs depend on the type of session but may include overhead projectors, electronic scanning board, videotape equipment, monitor, and tape player.
• Room temperature—between 72 and 78 degrees is desired.
• Seating arrangements—which can help the analyst control the session and manage multiple performers (see chapter 11).

Establish Active Leadership.   Establish active control of the session upon meeting the performer, using explicit or implicit control, or a combination. Explicit control includes welcoming or greeting the performer and introducing the session. Keeping the performer on track during the session is also a form of explicit control. Implicit control includes the use of body language subtleties and seating behavior.

State the Purpose of the Activity.   The analyst's inability to position the session as a professional, purposeful encounter negatively impacts the information acquired. After greeting the performer and establishing control, state the purpose of the activity. This includes a 2-minute (or so) explanation or "marketing statement" that describes the larger activity of system development. Why is the system being developed? Why are you interacting with performers? Who has authorized you to work with them? If a memo has gone out describing the activity, refer to it. After the general purpose has been stated, display the session report form and review the goals for this specific session. What needs to be accomplished by the end of the session? What is the most important type of information you are seeking? What type of session is it (interview, task analysis, etc.) and what will you be doing? This brief explanation enhances the psychological climate of the session and helps the performer view it as professional and task-oriented.

Guide the Performer Through the Session.   During the session the analyst acts as a facilitator, using nonverbal and verbal behavior to act in ways that enable session goals to be attained. Auger (1972) recommended the following tips that a facilitator could use to coax a session along:

- Stimulate discussion as responses cease.
- Balance the discussion if there are multiple performers participating to ensure each view is addressed.
- Keep the discussions on track.
- Ensure there is a positive, definable conclusion to the session.

Control the Time.   No one likes a meeting that runs considerably longer than expected. Analysts must watch the timetable carefully to ensure that session objectives can be met in the allotted time. We recommend that no session go longer than 90 minutes. Within this 90 minutes, human attention spans will wax and wane, possibly impacting the quality of the information elicited. One suggestion is to ensure the session starts on time and that no more than 5–10 minutes is spent on the greeting, introduction, and session goals. During the rest of the session the analyst should be aware that humans can attend actively for segments of 10–20 minutes. As the segment progresses, attention gradually starts to wane. Analysts can use this knowledge to help them structure the session, for example, starting a new primary question after approximately 10–20 minutes on one topic in an interview. By transitioning to a new topic, the performer's attention level can be raised. Finally, make every attempt to end on time, and if conversation is still productive, estimate the amount of time required to finish and give the performer a choice of continuing or re-scheduling.

Focus the Performer on the Appropriate Levels and Points. The analyst should know what type of information is sought and recognize if the information forthcoming will meet the session goals. For example, if the goal is to acquire in-depth information about how the performer diagnoses or troubleshoots a problem, the analyst should elicit deep level information about decision factors, alternatives, and heuristics. If the goal is to obtain a general view of the problem and its related concepts, the analyst should elicit broad, top level information. Once the analyst has identified the type of information desired, it should be pursued vigorously. If a question is not answered at the desired level of specificity, follow up with secondary, probing questions or restate the question to target the desired depth or focus.

Actively Summarize the Session and Debrief the Performer. At the close of a session, summarize accomplishments and debrief the performer. Summarizing session accomplishments enhances the chances that the performer leaves feeling that session objectives have been met and that the session was effective. Debriefing offers the analyst an opportunity to focus on what was accomplished, identify new areas to explore, and state the type of follow-up desired, if any (i.e., another session, material that might have been identified and requested).

Tape Sessions

Sessions may be audiotaped or videotaped to provide a record of session content, enable easier transcription and translation, and provide a means to session proceedings. Audio- and videotaping enable the analyst to concentrate more on the performer and on session management than on note taking. Both are appropriate in capturing session interactions and content, and enabling a more complete record from which to prepare the session report.

Videotaping. The prime advantage of the videotape is that information can be shared with other analysts and developers who might have been unable to attend the session. Furthermore, it enables the analyst to review not only verbal responses and interactions, but nonverbal ones as well. However, there are disadvantages to videotaping sessions. First, it can be distracting to the performers and thus may interfere with their ability to provide the information sought (McGraw & Harbison-Briggs, 1989). Some performers will become more introverted when a videotape is being made; others will become more extroverted—"hamming it up" for the tape. Second, videotaping requires some training and orientation on the part of the analyst and the performer. Third, transcribing a session from videotape is time consuming. In fact, Figure 3.8 illustrates our experience, which suggests that it takes up to three times as long as the session itself to review and document its content from a tape.

Audiotaping. While video is superior in its ability to capture nonverbal cues, interaction, and session context, it may not always be the best choice. Not only is it more expensive, it is also harder to transport and set up than audiotaping equipment. Furthermore, it is conspicuous and can distract performers (Hart, 1986). However, some people who are distracted by videotaping may not find audiotaping as intrusive (McGraw & Harbison-Briggs, 1989). We suggest using small, handheld voice-activated units that easily fit in briefcases or purses, are extremely portable, and are not foreboding when set in front of a performer or carried onto a factory floor.

Document Sessions

Each session may produce different output, depending on the goals of the session, the technique used, and tools (i.e., tapes) used. To compile

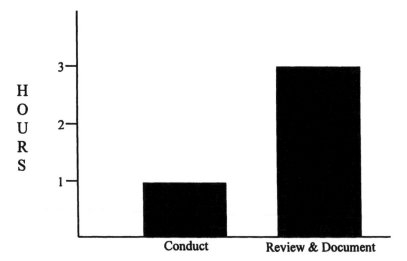

Time Requirements of Session Activities

FIG. 3.8. It can take up to three times the length of a session to review a videotape and document its contents.

documentation for a session, we suggest the following (McGraw & Seale, 1987; McGraw & Harbison-Briggs, 1989).

1. Keep informal notes during the session and use these to maintain a record of the general structures of the session. Refer to these notes when asking secondary questions and when summarizing the session. Keep these notes general; rely on audiotapes, videotapes, or a scribe (e.g., a junior analyst or other team member) to keep detailed notes. When you begin the documentation process, review session notes to establish a mental model for the session and recall information.

2. Attempt to review the notes and tapes as soon as possible—within 48 hours, if possible—to ensure that as little information as possible is lost.

3. Using the session report form, compose a general summary statement and record information that describes the performer's background.

4. On the form, note whether any documentation, work artifacts, videotapes, or audiotapes support this session.

5. Use an outline format to record main topics and subtopics described in the session. As the tape is played, identify the most relevant topic or subtopic and cluster information for later use. Summarize and restate the information succinctly as you create the body of the form. We rarely have a need to create a complete transcript of all sessions, but some sessions may be important enough to capture word-for-word responses. Scenarios or other specific information can be described or summarized in the report, and the point on the tape (i.e., the tape counter value) at which the information is presented can be documented in brackets for later review.

6. Mark any areas that are questionable, or items requiring performer review and comment using underline, italics, or highlighter.

7. Conclude the session report with any follow-up or action items that were discussed.

8. Submit the session report to the performer who participated in the session for review.

Figure 3.9 is an example of part of the body of a completed session form.

In addition to creating a session report, additional forms of session documentation exist, including:

- Intermediate representations, such as word-by-word transcripts from audiotapes or videotapes.
- Standard output for the technique used (i.e., task analysis sessions would be expected to produce a task/subtask hierarchy).
- Videotape or audiotape from the session, which should be labeled with the session number and stored in the reference center.

- New definitions and acronyms identified during the session, which should be included in revisions of the domain dictionary.

Handle Domain Materials

During a session, the performer may supply sample materials, job aids, work artifacts (e.g., reports, etc.), or screen prints to illustrate the topic being discussed. Materials received during a session should be logged on the session report form (for reference, and to establish an audit trail) and entered into the materials database. As the item is entered into the materials database, its title, session source, author, key words, summary, and

---

Session Report Summary

Joe Smith is the Division Commander for Emergency Medical Services. He is experienced in, and accountable for field services. He is responsible for central administration of the budget. Additionally, he supports and trains interns. He has 20 years of experience. He is certified at the EMT-P level.

Scene Assessment—Process and Primary Decision Factors

Scene assessment is presented in both BLS and ATLS training. In the basic course, about 20 minutes is spent on it. In the ATLS course, about an hour is spent on it. It is reinforced throughout the training program, however. It is taught in a practical way, supported with some pictures or slides. It is evaluated in a practical way, using scenario-driven evaluation The graphic illustrates a top level of the overall scene assessment process.

1. Receive the trauma call and gather enough information to begin forming assessments. Call information is received in two ways:

A. If in the station, you get a print out that may have some notes or information relevant to scene assessment. Examples include "known violent patient," "Tier 2 haz mat facility."
B. If in the unit (i.e., away from the station), you will receive verbal notification and information via the radio. The current equipment (for AA County) does not include a display in the unit, so you cannot see the notes or information if the call is dispatched in this manner.

2. As you prepare yourself and unit, begin considering the possible hazards you might face. Mentally prepare yourself, hypothesizing about known hazards in the area, or special considerations that will impact resource requirements, your plan for approaching the scene, etc. This is often based on your own personal experience, or notes received with the call.

Typical types of hazards you begin to consider enroute to the scene include the following:
- Hazardous materials—which facilities have them, which facilities are Tier 2 facilities, what materials are on site; sometimes a small, seemingly insignificant place like a local hardware store can have haz mat (e.g., pesticides)
- Violent patient (must be transferred over from the Police's notes)
- Special cases, high risk pregnancies
- Other medical hazards are difficult to transmit due to privacy act (AIDS, TB)
- Downed lines or electrical wires in the area
- Traffic situations
- The roadway enroute and at the site

FIG. 3.9. Partial body of a completed session report.

other meaningful information is documented as a part of the entry as illustrated earlier in the chapter.

Maintain a Domain Dictionary

A complex, large-scale software development project should maintain a domain dictionary to compile the domain's vocabulary and the special terminology pertinent to the project. Ask any student learning a second language, and you will find that a translation dictionary is a must! Likewise, to accurately define a reference architecture for a system, define a system's requirements, and work effectively with performers in the domain, analysts must understand the terminology and basic concepts.

During the session, analysts secure a definition of new terms used by the performer. As the analyst summarizes the session content, he or she compiles a list of new vocabulary and acronyms, along with their definitions or explanations. After the session is reviewed by the performer, changes are made to the definitions. Then the analyst compiles the terms, definitions, and acronyms, and notes the session in which they were defined (this can help sort out potential conflicts in definitions from different performers). Each new item is entered into the domain dictionary.

The domain dictionary is most effective when maintained electronically on a server. One person is responsible for updating and maintaining the file, but it can be downloaded, reviewed, and if desired, printed by everyone on the team. These suggestions may help compile and update a domain dictionary:

- During the planning stage, compile a list of basic terms and definitions. These may come from initial site visits or from documents the customer (or potential customer) provides. This "skeleton" domain dictionary is created using word processing software and is posted on the server.
- Determine who will update and maintain the file and ensure that team members understand how to submit new terms and acronyms.
- Individual analysts keep a running list of all new terms, key vocabulary, and acronyms as they conduct a session. This list is sent electronically to the person responsible for maintaining the domain dictionary.
- Send out electronic notices when the domain dictionary has been updated. This notice should include the terms and acronyms that were added. Team members can download a new version for their files.

Define Review Procedures

Accuracy and completeness of the information contained in session reports and output is critical. Establish review procedures throughout the require-

ments elicitation and analysis process to ensure that verification and refinement is an ongoing process. Reviews include performer review and
approval (or modification) of session reports, peer review of session reports
and outputs, internal management reviews, and external project reviews.
Chapter 12 suggests verification and validation procedures for reviewing
session reports.

Chapter 4

Selecting the Right Techniques

All professionals know that to complete a project successfully, they must use the right tools, at the right time, and use them correctly. A hammer is a very effective tool for most construction projects. However, if the hammer is the only tool used throughout the project, the result will not be fine workmanship. If our goal is to do something other than secure nails into woodwork, we would do better to select another tool lest we risk undesirable results.

Eliciting scenarios and requirements, translating them into design specifications, refining them, and modifying them for applications projects are distinct phases of a software development project. If we select a single technique as our "hammer" and use it in the same way throughout each project stage, the end result can be incomplete, inaccurate requirements and designs.

In this chapter we describe a well-accepted, comprehensive set of techniques developers can use to increase their ability to conduct domain analysis and elicit requirements that:

- Tap multiple types of critical knowledge in the domain.
- Enable performers to share their knowledge in the ways most comfortable to them.
- Help identify cognitive structures performers hold about the domain.
- Reflect performers' access to and use of information.
- Reveal decision making or problem solving processes and decision factors.

- Represent desired enhancement to the current process, or requirements for a new process.
- Represent processes within the context of the enterprise.
- Can be gathered in the most effective, timely manner.
- Enhance the ease of follow-up analysis.

In this chapter we introduce and describe the selected techniques. Next, we investigate variables that influence which techniques we select, including process phase, types of knowledge, elicitation requirements, and analyst capabilities. Finally, we examine each of these variables in detail, describing how they influence the selection of techniques, and the prime factors that should be considered.

Careful technique selection ensures that the analyst can select the right tool for the right need, and increases the probability of desirable results. Many problems related to requirements identification and initial design decisions can be traced to the selection of the wrong elicitation technique or the misuse of an appropriate technique. On small projects with one or two analysts and a cooperative customer, it is difficult to make too many "fatal" errors (i.e., errors that result in inaccuracies or grossly incomplete information). On larger, more complex projects, it is critical that techniques are selected carefully and that analysts and engineers are competent in their use. Failure may mean that (a) analysts must go back into the field and collect missing information, (b) information gathered could not be understood because it was not represented properly, or (c) information was acquired at the wrong level of detail.

The purpose of this chapter is to provide analysts with information that will enable them to consider all necessary variables when selecting elicitation techniques for use on a project. Specific goals include:

- Identify and differentiate primary elicitation techniques for domain analysis.
- Determine which techniques are most appropriate for specific process phases.
- Understand the importance of identifying knowledge type in selecting techniques.
- Recognize prevalent knowledge types in a domain.
- Understand the goals of a project, and carefully define the requirements of a knowledge acquisition or elicitation activity prior to selecting techniques.
- Recognize the capabilities and skills each technique requires, and be able to identify analysts' strengths prior to assigning tasks and techniques.

STANDARD SET OF ELICITATION TECHNIQUES

The field of all possible elicitation techniques that can be used during domain analysis is vast. It includes techniques borrowed from the field of communications (e.g., consensus decision making), psychology (e.g., construct analysis), instructional design (e.g., task analysis), journalism (e.g., interviews), and anthropology (e.g., observation). Hoffman (1987) analyzed knowledge elicitation from the perspective of experimental psychology and contended that all knowledge extraction methods fall into a handful of categories. Other authors uniformly describe the interview as nearly synonymous with knowledge elicitation.

Some techniques, such as the structured interview, are more effective with individuals; others can be used just as successfully in a group session. Some techniques, including the structured interview and process tracing, require special training or communication skills on the part of the knowledge engineer. Techniques such as decision process tracing require special hardware, software, or other props. Figure 4.1 illustrates a primary set of techniques for requirements elicitation. Each primary technique is described briefly in the sections that follow.

Scenario Generation and Use

Unfortunately, performers often have trouble defining what they want in a computer system. One technique to fuel requirements identification is to generate scenarios. Scenarios are "stories" or episodes depicting both

Decision Process Tracing

Scenario Analysis

Structured Interview

Interactive Observations

Concept Analysis

Work Process & Task Analysis

Group Techniques

FIG. 4.1. Primary SEP elicitation techniques.

normal and critical incidents that represent types of problems performers face (McGraw, 1994c). In the early stages of domain analysis, scenarios ease interdisciplinary communication (i.e., between performers in the domain, and computer designers and developers), and enable analysts to quickly identify preliminary system requirements (Hufnagel, Harbison, Doller, Silva, & Mettala, 1994). Later, scenarios help define responsibilities within a thread of execution and become test specifications for the system.

Analysts can generate scenarios by (a) eliciting them within the context of an interview or small group, (b) videotaping actual events within the workplace environment for later discussion and analysis, or (c) pulling case files for review and analysis. Additionally, a scenario may be acted out to illustrate how performers interact with each other and with the system (McGraw, 1994d). A scenario is recorded in a visual/verbal format, with narrative text and graphics (e.g., pictures or process flows) depicting activities within each segment of the scenario. After it has been captured and represented, a scenario can be used to draw out more in-depth information about the work processes, tasks, information use, and decision making it reveals (see chapter 5).

Interactive Observations and Site Visits

Interactive observation is one of a class of techniques that includes contextual inquiry (Holtzblatt & Jones, 1993), and is frequently used to obtain data by directly observing the activity under study. The purpose of interactive observation is to stimulate the revelation of the normal work process or task performance within the natural context of the job. It enables the analyst to observe and work with performers within the context of their job environment, and understand work activities from their perspective. It usually entails actual observation of work flow and its artifacts, followed by brief segments of questioning. The analyst uses observation output to understand better what is happening, how the performer is approaching the work, and what problems may be present.

Interactive observation requires site visits, which should be planned carefully. Sites should be selected to meet specific criteria, such as political importance, diversity of performers, adequate geographic representation, representation of both small and large, urban and rural sites, and so on. While on-site, analysts should have the opportunity to observe both expert and novice performers at work, ask questions at the completion of work episodes, and understand the context in which the work occurs (see chapter 6).

Structured Interviews and Surveys

Structured interviews and surveys enable analysts to gather a wide range of information in the early phases of a project. Later, these techniques can be used to gather specific details or to verify and validate information already gathered.

Both structured interviews and surveys present the performer with open and closed questions. Structured interviews are well-planned interactions in which the analyst asks specific questions and probes the performer after verbal and nonverbal responses. Interview session goals define question sets and help identify sample questions; the analyst uses questioning and listening techniques to elicit responses and seek the level of detail required. This comprehensive, flexible technique can be used at any point in the requirements gathering process or during design feedback sessions. It can be used alone or in combination with observation sessions, task analyses, and concept analyses, or after a decision process tracing session.

Surveys provide the analyst with a vehicle to ask the same types of questions of a number of performers without the logistical problems of individual interviews. The resulting information presents responses across a range of performer types, roles, and locations. However, surveys provide only verbal (as opposed to nonverbal) feedback, and do not offer the opportunity to probe for specific details or examples (see chapter 7).

Work Process Analysis

Work process analysis enables analysts to understand the top-level functions or responsibilities of a person's job. It helps us "get our arms around" the total activity of a job and the major accomplishments it represents. To reveal work processes, analysts investigate a number of issues. These include job responsibilities, sources of work, types of output, critical success factors, means of measurement, impact of a single work process on the total job, and motivation. Consequently, what we learn during interactive observation fuels the work process analysis. By obtaining detailed knowledge about the structure and flow of the work process, designers can define and structure systems that support and extend the work.

The output of a work process analysis can take many forms, depending on the goals of the project. Most often, these are variations of tables and flowcharts that depict the issues investigated (see chapter 8).

Task Analysis

Task analysis is a methodological family of techniques that can be used to describe the functions a human performs and to determine the relation of each task to the overall work process (Shannon, 1980). It enables the decomposition of each work process into its tasks. Each task is then analyzed to determine task goals, operations, and requirements. The purpose of a task analysis is to classify and decompose tasks involved in carrying out a particular work process and meeting performance goals.

The output from a task analysis is a task model. Task models can describe tasks and subtasks and the skills, knowledge, abilities, information, and materials required. Analysts may use any of a variety of task model representations, including hierarchical diagrams, task/subtask matrices, and templates. Analysts work from these representations to establish prerequisites and functional, informational, and usability requirements for components and devices (see chapter 8).

Concept Generation, Organization, and Analysis

Concept-related techniques help analysts determine the classification and categorization techniques and schemes performers use. First, the primary concepts and attributes that differentiate them are identified or defined. Next, the relationships among these concepts and attributes are discerned.

Concept analysis involves eliciting the names and definitions of primary concepts in the domain, and an initial categorization and classification scheme that can be used to organize and understand them. Analysts prompt performers to identify and label items, group items, label groups of items, define items, and identify a common factor among items. Concept analysis may also involve comparing and contrasting concepts and conceptual groups to isolate the salient features of primary concepts, and to be able to discriminate between similar concepts.

The output of concept analysis includes categories of concepts in the form of organizational charts and diagrams, in addition to frame-like structures in which various slots provide a structure in which a concept definition can be presented (see chapter 9).

Decision Process Tracing

Decision process tracing involves asking performers to report on or demonstrate their decision-making process for a specific problem. The analyst develops a structure or framework that can be used to represent the information, actions, alternatives, and decision rules the performer is using (Svenson, 1979). This technique has many variations, including think-aloud protocol construction and retrospective, or taped analysis of protocols. Decision process tracing is effective in eliciting routine procedures, facts, or heuristics; it is especially useful in helping reveal hypotheses the performer considers and the decision factors and reasoning the performer applies. Analysts ask the performer to (a) complete a task and explain reasoning involved or (b) complete a task that is videotaped, then report on (and answer questions to explain) the process and decision making factors and heuristics they used.

The output of decision process tracing may include flowcharts, decision–action diagrams, rules, or constraints within frames or objects (see chapter 10).

Group Techniques

Any of the aforementioned techniques may be adapted for use with multiple performers. However, special techniques may be required when performers are consulted in a small group situation, including:

- Brainstorming and mind mapping—These techniques help analysts guide performers to "break loose" from obvious, conventional solutions to complex problems. They are designed to stimulate thinking, generate ideas, prohibit immediate criticism, and reduce discussion-inhibiting comments.
- Focus group design and management—Focus groups provide a forum in which analysts can lead a small group of performers through a series of questions that have a common focus. The goal is to elicit feedback that can be used to refine the ideas presented. These techniques often include the demonstration of a prototype of a product, a set of storyboards, or sample screens from an application.
- Consensus decision making and nominal group techniques—The goal of both of these techniques is to isolate the best solution or set of solutions to a problem, and identify which solutions are unacceptable by the group. Consensus decision making involves presenting a problem or question to the group and encouraging each member to vote on alternative solutions or answers. Voting is verbal and takes place in rounds, enabling each member of the group to make his or her views known, regardless of rank or status. The nominal group technique reduces the negative effects that may be triggered by face-to-face interaction. In this technique the group co-exists, but its members are allowed to vote and comment anonymously (see chapter 11).

VARIABLES INFLUENCING SELECTION

Many variables will influence the selection of specific requirements elicitation techniques and the degree of structure with which they are applied. It may be advantageous to use a specific technique if it (a) enables meeting established program goals, (b) makes efficient use of the analyst's and performer's time, or (c) is less threatening to the performer.

The analyst must weigh advantages and disadvantages when selecting a technique. For example, the interview may yield a tremendous amount of data. Yet it is often time consuming, requires that the analyst have basic interviewing skills such as active listening, question development, and interview management, and may result in copious but surface-level knowledge.

Determinants for the selection of requirements elicitation techniques include the following, each of which is discussed in the sections that follow:

- Process phase.
- Types of knowledge predominant in the domain.
- Elicitation requirements.
- Attributes and characteristics of techniques.
- Analyst capabilities and strengths.

Process Phase

As you progress through phases of a software project, your goals change to reflect the accomplishments that must take place before you move into subsequent phases. The requirements elicitation and domain analysis techniques you select should help you meet the goals of the process phases in which you are working (McGraw & Seale, 1987). In this case the phases we examine are primarily those that occur during project planning, requirements elicitation, and requirements analysis. However, many of the techniques we propose are also quite effective in the design and development activities. Some of them (primarily the group techniques and interviews/surveys) can also be applied during evaluation activities. Different techniques are appropriate during different phases of activities, as Table 4.1 indicates.

Planning

The first major set of activities that occurs is planning. This phase includes preliminary investigation and problem definition activities (Senn, 1989). Preliminary investigation and problem definition requires that we quickly understand, at a general level, what performers and their managers want. However, few groups are able to clearly explicate what they need without some assistance in problem identification or clarification.

We have successfully used requirements elicitation techniques to assist in identifying and clarifying problems and possible solutions. As Table 4.1 shows, interactive observations and site visits are good choices when potential customers are unsure of the problem and their needs, or when a preliminary investigation is required to help refine the problem definition. If an extensive investigation is unnecessary, group techniques such as focus

TABLE 4.1
Selecting Different Techniques According to Process Phase and Goals

Techniques	Planning	Domain Familiarization	Generalizing Work Process & Identifying User Requirements	Decomposing Tasks	Input, Output, & Information Requirements	Understanding Problem Solving Activities	Designing Screens, Objects, & Models	Refining & Evaluating Requirements, Designs, & Functionality
Interactive Observations & Site Visits	●	●	●					
Work Process Analysis			●					
Scenario generation & analysis		●	●	●	●	●	●	
Interviews & Surveys			●		●		●	●
Task Analysis				●	●		●	●
Concept Generation, Sorting, and Definition		●				●	●	
Document & Artifact Reviews & Frameworks		●				●	●	
Decision Process Tracing		●	●	●	●	●	●	
Focus Groups	●		●		●			●
Brainstorming & Mind Mapping	●	●				●	●	
Consensus Decision Making					●	●	●	●
Nominal Group					●	●		●

Phase Goals

groups and brainstorming or mind mapping can be effective. These techniques can help reveal the organization's critical success factors, problems preventing goal achievement, and desired innovations, process changes, or supporting systems.

Domain Familiarization

Domain familiarization is the process of becoming familiar with aspects of the domain for which a new system (people- or technology-based) will be developed (McGraw & Harbison-Briggs, 1989). Most developers agree that it is useful to invest time in gaining an initial understanding of the domain before beginning full scale system development activities. Because most projects are staffed with developers whose expertise is in systems design and development, the importance of this phase cannot be minimized. Domain familiarization activities enable the system development team to do the following:

- Develop an initial mental model of the domain from the performer's perspective.
- Learn key vocabulary used to describe activities in the domain.
- Identify the general work context for activities a system will support.
- Begin building team-oriented relationships with members of the customer's organization.

The most effective techniques for meeting the goals of domain familiarization are interactive observations and site visits (if not already conducted). Seeing *is* believing, and helps the team become familiar with the domain quickly. An alternative to site visits could include asking a performer to videotape activities on site, then showing the tape (complete with the performer's description of activities) to the development team.

Document reviews and frameworks can also help the developer build mental models and learn key vocabulary for the domain. This includes studying manuals and training documents, and compiling a document framework that decomposes important content for easy reference and use by the development team.

Generalizing Work Process and Identifying User Requirements

As discussed previously, the goals of a work process are to identify and define a performer's major job functions. Thus, a work process analysis enables analysts to begin to understand the problem setting and the performers involved. This process phase helps developers answer the following questions:

- What is being done?

- How frequently does it occur?
- What is the volume?
- Who does it?

Numerous techniques are appropriate during this process phase. Their selection depends on whether they have been used in a previous phase, and on the amount of time and money available for activity in this phase. Interactive observations and site visits must be completed to understand the work process. Selected interviews can be combined with surveys, enabling performers to respond to questions about their primary work processes, work process goals, volume, and frequency. Selected performers complete the surveys, which are sent to analysts for review. Later, analysts conduct follow-up interviews (in person, via phone, or videoconference) with each performer, to ask questions about information provided on the survey. Information elicited is used to begin the work process analysis.

During this phase, analysts also should review manuals, job aids, documents, and work artifacts, such as reports or products created as a result of the work process. Finally, focus groups may be held to present the initial understanding of the work process and to refine it prior to task analysis activities.

Decomposing Tasks

Analysts decompose tasks to describe the functions a person performs and determine the relation of each task to the overall work process. Task analysis activities enable analysts to move beyond simple identification or definition of required tasks. The purpose of this activity is to answer the following questions:

- How is work being done?
- How well is it being done?
- What are the performance problems and where do they occur?

Techniques that are effective during this process phase include document reviews and frameworks, interviews and surveys, and task analysis activities. Documents may also reveal steps performers should take to complete tasks, materials used, acceptable performance goals, and so on. Interviews can help elicit more complete information about tasks already observed. Task-specific surveys can help analysts reveal information such as performers who complete specific tasks, tasks rated as hard to do, tasks rated as hard to learn (McGraw, 1994b), and so on.

Outputs of task analysis include task hierarchies, timelines, and task matrices. These devices help analysts understand the components and requirements of tasks, which can result in more complete definition of requirements for functionality and system design.

Input, Output, and Information Requirements

Once we have identified tasks and task steps, we can define input, output, and information requirements associated with each task. This process phase helps us answer the following questions:

- What is required to begin tasks?
- In what format is information input?
- How is information processed during a task?
- What is output, and in what format, at the completion of the task?
- To whom does the output go?

Activities appropriate for defining and understanding information requirements include reviewing documents and work artifacts, and reviewing or completing task analysis output. Conducting decision process tracing sessions and follow-up interviews or surveys may also be beneficial in revealing information use.

Note that during this phase the primary focus is on reviewing and reusing information already gathered during other process phases. For example, analysts review task analysis output, such as a hierarchical task graphic and a tabular task matrix to identify information that is an input for a task. Additionally, they attempt to discover how information is manipulated or processed during task completion and identify the output of each task. After they review the output from decision process tracing sessions they identify the information required to make decisions. If the required information is not revealed, analysts may conduct follow-up interviews. To enable them to gather information from a large number of performers, analysts may use a survey to identify information used to complete tasks, the importance of different kinds of input, and problems relating to task input. Finally, analysts may review existing documents (e.g., data dictionaries and process diagrams) to identify information provided as input, processes supported by existing systems, and output from system-supported processes.

Understanding Problem-Solving Activities

Decision making and problem solving require functional understanding of concepts in the domain, and the ability to organize and apply concepts and heuristics effectively. Questions answered during this phase include:

- Which tasks involve decision making or problem solving?
- What type of decision making occurs (forward vs. backward chaining)?

- What factors are considered?
- How do you know when the problem is solved and the correct decision made?
- How do experts and novices differ in solving domain problems?

The most effective techniques during this phase include concept generation and sorting, and decision process tracing. Concept generation and sorting can begin with a review of task analysis output or scenarios that have been gathered and analyzed. Next, analysts work with performers to define concepts and understand how they are retrieved and applied during decision making. To do so, analysts may also create and review videotaped artifacts of completed problems, decisions, or scenarios to better understand how performers solve problems.

If the process being examined involves an extensive number of performers, or if it is handled by a group, analysts may also conduct sessions such as brainstorming, mind mapping, or consensus decision making. These techniques help us ensure that the solutions, decisions, heuristics, and factors identified are acceptable to a wide range of performers and are not subject to individual biases.

Designing Screens, Objects, and Models

A great deal of information that can be used to create screens, objects, and models can be obtained by reviewing the output of techniques already applied. Questions answered during this phase include:

- What functions must the screens support?
- Do the screens match performer mental models and meet performers' needs without extensive training and support?
- What is the best way to present functions, icons, and concepts?
- What objects, models, and components are required to support system functionality?

A number of techniques can provide input to this process phase. Analysts may want to start by reviewing current scenarios and sketching primary screens for a proof-of-concept or prototype system. Further definition of initial screens can be obtained by reviewing task analysis output and ensuring that the functions and organization of the screens will support the required tasks. Screens should be designed to alleviate problems identified during the task analysis and to support performers' decision making. Existing documents that support system design, such as data dictionaries and process diagrams for legacy systems, should be reviewed during this phase.

Techniques that could be conducted anew in this phase include concept and object generation, definition, sorting, and modeling. These activities require the use of a small team consisting of analysts, developers, and performer representatives. Alternatively, analysts and developers may lead a small group of selected performers through these activities using consensus decision making.

Refining and Evaluating Requirements, Designs, and Functionality

Many of the elicitation techniques can be modified for use during the evaluation and refinement of requirements and designs. At this phase the following questions are answered:

- Does the proof-of-concept or prototype reflect the desired features?
- Does the system prototype work properly to perform its specified function?
- Are the functions depicted feasible, from technical and cost perspectives?

Both individual and group techniques can be used effectively during this process phase. Scenarios, which have already been elicited and analyzed, can be used to evaluate the completeness of preliminary requirements, designs, and prototype functionality. Interviews, surveys, and questionnaires can be used to assess how performers evaluate preliminary requirements and designs as they interact with the prototype versions of the system. Alternatively, analysts may present the prototype system to small groups of performers during focus groups. At this stage the purpose of the focus group is to document performers' reactions and get answers to specific questions analysts have about the preliminary requirements and designs. To refine requirements and designs, analysts also may use group techniques, including brainstorming and mind mapping, consensus decision making, or the nominal group technique.

Type of Knowledge

The second major factor analysts should consider when selecting techniques is type of domain knowledge. Gammack and Young (1985) contended that knowledge engineers and analysts tend to select techniques based on how difficult they are, rather than investigating which ones are the most appropriate for the situation. They conclude that a frequent problem in the elicitation and analysis of domain knowledge is that selected techniques are not appropriate for the type and nature of knowledge in the domain. McGraw & Harbison-Briggs (1989) believed that part of the test of "appropriateness" is determining the primary type or types of knowledge prevalent in the domain and ensuring that the selected technique can help the analyst elicit and translate it with the greatest ease. Performers may use one, several, or all of these knowledge types to function effectively in the domain; however, one knowledge type is usually primary.

Analysts should investigate the domain and characterize the prevalent knowledge types to enable them to select appropriate elicitation techniques. Figure 4.2 illustrates one way to categorize the basic types of knowledge. The sections that follow briefly review each knowledge type.

Declarative Knowledge

Declarative knowledge represents surface-level knowledge and includes things we are instantaneously aware of knowing. Its contents are present in short-term memory, requiring little or no active retrieval. It includes the names of objects in a domain, descriptors that identify common items, and definitions of their use. It also includes conscious general heuristics— top-level rules or facts pertaining to the domain.

Declarative knowledge is very useful in the initial stages of requirements elicitation—generating scenarios, comprehending the overall work process, understanding the basic terms and performer roles. However, it may *not*

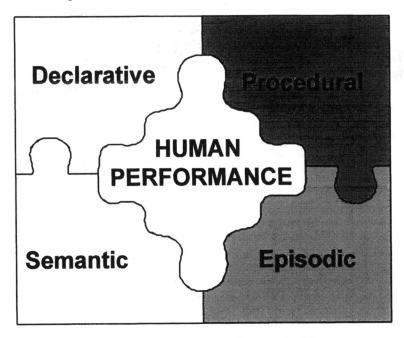

FIG. 4.2. Critical knowledge types analysts must be able to tap.

accurately depict the cognitive foundations and concepts a performer uses to relate or organize the information in a meaningful fashion (McGraw & Seale, 1987).

Procedural Knowledge

Procedural knowledge includes the skills an individual knows how to perform. As such, it often involves an automatic response to stimuli and may be reactionary in nature. Examples of procedural knowledge include learned psychomotor skills, such as executing a sharp turn while riding a bicycle. However, procedural knowledge is not simply related to gross motor behaviors. Researchers have noted that knowledge of one's own native language is highly procedural (Anderson, 1980). Most of us are effective, facile speakers of our native tongue. To be able to produce language effectively, we employ thousands of rules about how words are pronounced, which words we should use to encode an idea, how words should be linked, and how the parts of speech work together within the world of grammar. Furthermore, if we interrupted an expert speaker to inquire about these rules he or she would find it difficult to state these rules. Figure 4.3 provides some examples of procedural knowledge.

Rules for applying this type of knowledge are deeply embedded and linked sequentially. The evidence of a performer's ability to apply procedural knowledge is often called skills. As a skilled performer completes one step in a procedure, it triggers the performance of the next step. The steps may be so highly compiled that they are difficult for the performer to isolate and identify, much less discuss (McGraw & Seale, 1987). As Anderson (1980) explained it, when the same knowledge is used over and over in a procedure, we lose access to it. Consequently, our ability to report it diminishes.

Ask any expert pilot how he or she completes part of the process of take-off and you quickly realize that the steps are so linked and deeply compiled that the pilot may find it difficult to know where to begin. Another example is our ability to recall and report a frequently-called telephone number. When we first memorize a telephone number we can verbally report it *and* dial it—moving our fingers in a pattern across the phone pad—with equal ease. After repeatedly dialing the number we actually memorize the pattern and may be able to enter the numbers quickly without more than a glance at the phone pad. At this point we begin to experience difficulty reporting the number. We have dialed it so many times that we have compiled it as the pattern. To verbally report it we must first decompile the pattern into its parts.

The primary distinction between procedural and declarative knowledge is that performers can easily tell us about declarative knowledge. They find it much more difficult to describe procedural knowledge. Consider how

FIG. 4.3. Examples of domains with prevalent procedural knowledge.

difficult it would be for most of us to describe how to tie a shoe accurately, without actually doing it.

Semantic Knowledge

Semantic knowledge represents one of the two theoretical types of long-term memory. Semantic knowledge reflects cognitive structure, organization, and mental representation. As Fig. 4.4 indicates, researchers (Tulving, 1972, 1983) believe semantic knowledge is organized knowledge about the following:

- Words and other verbal symbols.
- Word/symbol meanings and usage rules.

1. Words and other verbal symbols

2. Words meanings and usage rules

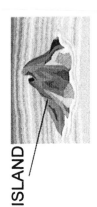

ISLAND

3. Word referents and interrelationships

I met Jane. She was nice.

4. Algorithms for manipulating symbols, concepts, relations

COW   COWS   cattle

FIG. 4.4. Examples of semantic knowledge.

- Word/symbol referents and interrelationships.
- Algorithms for manipulating symbols, concepts, and relations.

Semantic knowledge includes memories for vocabulary, concepts, facts, definitions, and the relationships among facts. Consequently, it is of primary importance to the knowledge engineer or analyst (McGraw & Seale, 1987). In fact, semantic knowledge will determine whether an information system actually supports the *real* work a performer must do in the given domain.

Episodic Knowledge

Episodic knowledge is autobiographical, experiential information that performers group or chunk according to episodes. Episodic knowledge is theorized to reside in long-term memory. It contains information about "temporally dated episodes or events (i.e., what happened and when it occurred) and time-space relations among these events" (Tulving, 1972, p. 385). Best (1986) contended it is organized by time and place of occurrence, and noted that it is described in terms of perceptual characteristics. For example, in response to a question about how diagnoses are made during troubleshooting, an expert performer might begin to describe a particular event that "comes to mind." Embedded within the event is a description of the attributes of the situation—the setting, the problem, how devices and components looked, smelled, responded, and sounded. Additionally, the performer would probably continue describing the episode, recalling the tests that were run, results that confirmed a hypothesis, and so on.

Wallace (1972) presented a detailed analysis of a type of episodic knowledge that most of us use on a daily basis—driving to work. Although we use this knowledge daily, we may find it very difficult to abstract any discrete rules about how we drive to work, how we drive the car, our speed per stretch of road, lane choices we make, and so on. Some of us may not even be aware of the names of roads we take. The information is ingrained and chunked episodically (McGraw & Harbison-Briggs, 1989). Consequently, many of us have experienced the phenomenon of arriving at work only to realize we cannot recall our actions during certain portions of the trip.

Because episodic knowledge is highly compiled and autobiographical, it is one of the most difficult types of knowledge to extract and dissect. Tapping episodic knowledge often necessitates the use of task analysis and protocol analysis techniques, including the analysis of videotaped episodes (McGraw & Seale, 1987).

Matching Techniques to Knowledge Types

Table 4.2 illustrates a suggested mapping of knowledge type to elicitation technique. If the activity is "identify general information and heuristics available on a conscious level," for example, the analyst is seeking surface

TABLE 4.2
Correlating Knowledge Type and Elicitation Technique

Requirements Elicitation Target	Type of Knowledge	Suggested Technique
General information about or used by performers—facts, names, characteristics, general work processes, critical success factors, etc.	Declarative knowledge	Interactive observation Structured interview
Routine procedures or tasks completed by performers	Procedural knowledge	Structured interview Interactive observation Work process analysis Task analysis
Major concepts, concept organization, vocabulary used in the domain	Semantic knowledge	Concept generation Concept sorting Document frameworks
Decision making procedures and unconscious heuristics	Semantic knowledge	Decision process tracing Scenario analysis Task analysis
Problem recognition cues Analogical problem solving heuristics	Episodic knowledge	Scenario analysis Process tracing

level, declarative knowledge. Declarative knowledge is available in the short-term memory, which enhances the ease with which we can talk about it. Almost any technique can be used to elicit knowledge we can easily talk about, but recommended techniques for this situation include the structured interview and interactive observation.

If the analyst is seeking to identify "analogical problem solving heuristics," the interview would be ineffective. Analogical problem solving requires that the performer view a new situation, make analogies to problems already solved, and apply or modify potential solutions for the new problem situation. This type of knowledge and thinking requires episodic knowledge, which is very difficult to relate verbally. In this case, techniques such as process tracing on scenarios or selected cases would be more appropriate.

Elicitation Requirements

Another consideration for the selection of techniques is the project's requirements and their effect on the level of detail the elicitation activity must yield.

What are the requirements for elicitation? What must the elicitation activity produce? What level of detail is expected? Analysts should define an approach and select techniques that will yield the level of detail and output appropriate for the project. For example, on one project you may be asked to conduct complete requirements identification and analysis to define and build a system. This type of project requires that analysts conduct planning, requirements identification, and requirements analysis. In this case they might use the following techniques:

1. Interactive observation and site visits to become familiar with the domain and to plan the elicitation activities.
2. Document analysis and the construction of document frameworks that define key work processes or job functions, vocabulary/terminology, and primary information sources.
3. Work process and/or scenario analysis, followed by task analysis, to identify the work processes performed. These are then decomposed into tasks to reveal input, output, processing, and information requirements.
4. Concept analysis on concepts and objects identified during task analysis.
5. Decision making analysis to identify heuristics and problem solving approaches associated with the tasks.
6. Brainstorming with small groups of performers to identify primary problems and help define the "to be" or revised work process.
7. Focus groups with small groups of performers to present screens for a proof-of-concept that will enable them to refine requirements.

On the other hand, not all activities result in the development of a proof-of-concept, a prototype, or a requirements document. Some projects go no further than the planning phase discussed earlier. In cases such as this, you might be asked to help a customer determine a particular approach to a performance problem—not to develop full requirements. To meet this goal, you might conduct only the following types of sessions:

1. Focus groups with managers to identify the organization's critical success factors and mission, and to understand their goals for the project.
2. Interactive observation and site visits to become familiar with the domain.
3. Scenario and work process analysis to identify performance goals and problems.
4. Brainstorming and consensus decision-making groups with performers to identify solutions to recognized problems.

After these sessions have been conducted you might simply prepare a report or make a presentation that suggests alternatives and the most effective approach. In this case, the level of detail is much less than in the previous example. Had you conducted task and decision-making analyses you would have exceeded the goals of the project.

Attributes and Characteristics of Techniques

In selecting the techniques to use for a project, analysts must also consider the attributes and characteristics of each technique. They vary in their effectiveness in specific situations, the ease with which they can be conducted, and the ease with which their output is transcribed or translated. Hoffman (1987) proposed that analysts use a set of desirable criteria to help compare and select techniques. Techniques may be rated according to the following attributes and characteristics:

- Task simplicity—the ease with which the analyst can use a technique.
- Material simplicity—the materials or props required.
- Task brevity—the minimum length of a session in which a technique is applied.
- Task flexibility—the extent to which a technique may be modified and reused.
- Task reality—the degree to which the technique represents the real world.
- Translation ease—the effort required to translate and analyze information elicited using the technique.
- Data validity—the extent to which analysts can feel confident that the data elicited with a technique is valid.
- Technique efficiency—the amount of information elicited in relation to the time expended.

Table 4.3 presents questions that can be asked to determine if a technique meets each of these characteristics.

By rating techniques according to these attributes, analysts can develop a comparison table that can be used to help select a core set of techniques, or pick one for a particular session. Figure 4.5 is an example of a job aid an analyst can use to rate techniques. Enter the application or project name, then fill in the techniques under consideration. Rate each technique according to the set of criteria deemed critical. Add up the affirmative marks and compare the ratings among techniques.

Analyst Capabilities and Strengths

The selection of a technique should also depend on the analyst's skills. Although all analysts can be trained to use all of the techniques, some

TABLE 4.3
Sample Questions and Criteria for Rating Techniques

Criteria	Sample Question(s)
Task simplicity	How much training is required prior to the effective use of this technique?
	How easy is the technique to use?
Material simplicity	How easy is it to use the materials or "props" this technique requires?
	What is the ease and cost of acquiring the required materials?
Task brevity	How much time is required to complete a session with this technique?
	How much follow-up, transcription, or analysis time is required?
Task flexibility	Is the technique adaptable to new situations?
	Can the technique be modified to meet the needs of a specific session or performer?
	Can the technique be used with a variety of materials?
Task reality	How closely does the task required of the performer (in the session) match the real world?
Translation ease	How easy is it to translate or represent information acquired using this technique into a usable format?
Data validity	How valid is the data acquired using this technique?

individuals will possess personalities, innate capabilities, and skills that enable them to apply some techniques more effectively.

Training, which can improve *skills*, does not impact personality or innate capabilities. Factors such as an analyst's ability to facilitate, communicate, lead sessions, and translate and organize session output have proven to be critical in the effective use of techniques. Table 4.4 illustrates the impact of these factors on different techniques. In this table, H indicates a high degree of skill required, M indicates a moderate degree of skill, and L indicates a low degree of skill required. These requirements should be considered when selecting techniques.

The sections that follow provide an explanation of the attributes and capabilities that have the strongest potential impact on the effectiveness with which an analyst selects and applies a technique.

Analyst as Facilitator

Analysts act as facilitators as they enable session goals to be attained. Some techniques, such as the interview, concept analysis, task analysis, and group sessions require more facilitation than others. Facilitation requires that the analyst use nonverbal and verbal behaviors to stimulate the discussion, balance the discussion if there is more than one performer, keep discussions on track, break up "hot" controversies, and manage time effectively (Auger, 1972).

Technique	Task Simplicity	Material Simplicity	Task Brevity	Task Flexibility	Task Reality	Translation Ease	Data Validity
Structured Interview	○	●	●	●	○	○	○
Interactive observation	●	○	○	●	●	●	●
Work process analysis	○	●	○	○	●	●	●
Scenario generation & analysis	●	●	●	●	●	○	●
Task analysis	○	●	○	○	●	●	●
Decision process tracing	●	○	○	●	●	○	○
Special group techniques	○	●	○	●	○	○	○
Concept generation and analysis	○	●	○	●	○	●	○

● = "yes"
○ = "no"

FIG. 4.5. Example of a job aid for selecting techniques for a specific application.

TABLE 4.4
Facilitation, Communication, Leadership, and
Translation Skill Requirements Vary by Technique

Technique	Facilitation Skill	Communication Skill	Leadership Skill	Translation & Organization Skill
Interactive Observations & Site Visits	L	M	M	M
Work Process Analysis	M	H	M	M
Scenario generation & analysis	M	H	H	H
Interviews & Surveys	H	H	H	H
Task Analysis	H	H	M	H
Concept Generation, Sorting, and Definition	H	M	H	H
Document & Artifact Reviews & Frameworks	L	L	L	M
Decision Process Tracing	M	M	M	H
Focus Groups	H	H	H	H
Brainstorming & Mind Mapping	H	M	H	H
Consensus Decision Making	H	M	H	M
Nominal Group	H	M	H	M

Analyst as Leader

Some techniques, such as the interview, scenario generation, concept generation, and group techniques, are more open or unstructured than others. These techniques require that the analyst act more in a leadership role than do techniques that are structured by the environment, such as interactive observation. Analysts vary in their leadership skill and style. Careful attention should be paid to ensure that analysts with poorly developed leadership skills are not asked to rely on techniques that require a high degree of leadership skills.

Two areas that are particularly relevant to leadership of requirements elicitation sessions are goal setting and achievement, and interpersonal management (Ross, 1983).

Setting and Achieving Goals. To achieve goals set for sessions, analysts must be capable of contributing and evaluating ideas, isolating issues, synthesizing ideas of others, and gaining consensus.

First, analysts must ensure that each participant has the opportunity to contribute ideas to the session. Once contributed, ideas can be evaluated through effective questioning and comparison to other ideas. Evaluation occurs even in sessions that involve only one performer. In this case the *analyst* must carefully consider a response and ask questions that clarify incongruous responses. Analysts also must force the performer to expand on seemingly conflicting comments or global, meaningless statements (see "questioning techniques," chapter 7).

Second, the analyst must help isolate the real issues. In some sessions performers may simply know what the key issues are and assume the analyst understands them. In others, performers may not be cognizant of key issues. In either case the analyst must listen carefully for statements that harbor issues, and must have a firm enough grasp on key concepts to recognize the issues and target them through comments and questions.

Third, analysts must be able to synthesize the ideas of others. This includes recognizing the meanings inherent in a statement and the ability to extract that meaning, synthesize it with other ideas, and restate it for evaluation.

Finally, analysts must be able to gain consensus. When disagreement occurs, they must be able to recognize the source of disagreement and determine if consensus is necessary. If so, they guide performers through activities designed to seek or build consensus.

Managing Interpersonal Interactions. To manage interpersonal interactions within a session, analysts must be able to resolve conflicts and control emotions, set the communications climate, regulate contributions, and promote interaction.

Conflict management is particularly important in sessions that involve multiple performers. Analysts may vary in their ability to manage conflicts. Figure 4.6 illustrates five recognized conflict management styles (Schermerhorn, Hunt, & Osborn, 1991). These styles vary along two scales—*cooperativeness*, the inclination to satisfy the other party's concern, and *assertiveness*, the inclination to satisfy personal concerns.

Smoothing or accommodation involves letting the other's wishes rule and smoothing over differences to maintain superficial harmony. At the other end of the cooperativeness scale is avoidance, in which we downplay disagreement, fail to participate in the situation, and/or stay neutral at all costs. Problem solving is the most effective conflict management style within sessions. With it we seek true satisfaction of everyone's concerns by working through differences and finding solutions in which everyone gains. The compromise style of conflict management involves working toward partial satisfaction of everyone's concerns and seeking "acceptable" rather than optimal solutions. Four of the five conflict-management styles shown have some potential value to analysts; however, conflict avoidance should never be used in elicitation sessions.

In addition to being able to manage conflicts, analysts must be able to set the communications climate of a session to stimulate responses that

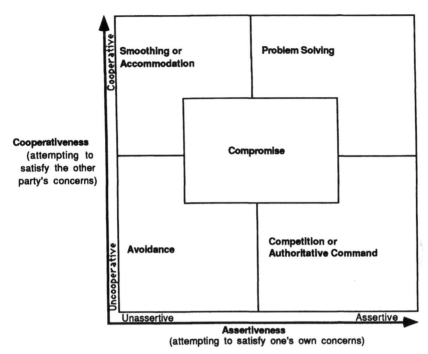

FIG. 4.6. Conflict management grid (Schermerhorn, Hunt, & Osborn, Managing Organizational Behavior, © 1991. Reprinted by permission).

are accurate and complete. The following attitudes help create the desired communications climate:

- Team-oriented, in which each participant works with the other to accomplish session goals.
- Open and accepting, so that participants feel they can be honest.
- Purposeful and goal-oriented, so that participants recognize there are session goals and that the session is managed to ensure their achievement.

Finally, analysts managing sessions must be able to regulate contributions and promote interactions of session participants. Even if there is only a single participant, the analyst must be able to apply effective leadership styles to manage the session flow and keep it on track. Figure 4.7 depicts a range of leadership styles, from little or no control to full control. Laissez-faire style represents an "anything goes" situation. It is effective only in situations in which the analyst is a bystander, such as site visits, or viewing or videotaping activities during scenario generation. Nondirective and permissive styles represent the use of some control. However, neither is con-

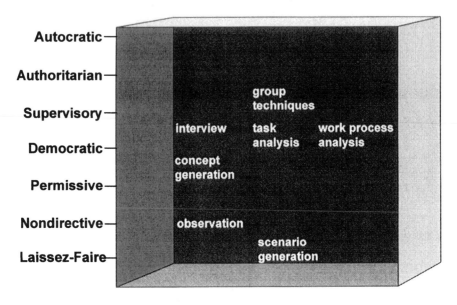

FIG. 4.7. Leadership styles required vary by technique, but most techniques require democratic or supervisory interaction.

sidered directive enough to manage techniques other than interactive ob-servation sessions. Most techniques, including the interview, work process and task analysis, concept generation, and group techniques require man-agement styles in the democratic and supervisory range. Strict authoritarian and autocratic styles are inappropriate for most situations because they have a negative impact on interpersonal relationships and communication. However, analysts may need to move into this area during group sessions to control emotions or resolve heated conflicts.

Analyst as Communicator

Many variables affect one's ability to communicate effectively in sessions, including interaction or communication style, questioning and listening skills, and the ability to use the vocabulary of the domain.

Interaction or Communication Style. Perhaps no variable is as pervasive in its influence on communication as the interaction style of the session facilitator or analyst. Interaction or communication style includes the sig-nals provided to help process, interpret, filter, or understand literal mean-ing (Norton, 1983). One mechanism used to do so is the feedback provided during a session. In a session with high communication requirements (such as the interview), the analyst and performer interact using feedback that

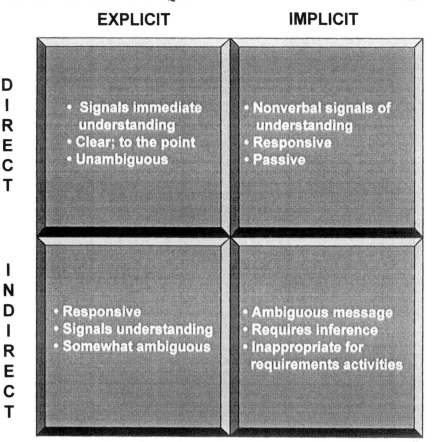

FIG. 4.8. Feedback matrix (adapted from R. Norton, Communicator Style, p. 179).

is either explicit or implicit, and direct or indirect. Figure 4.8 depicts a matrix of the four styles of communication feedback.

A highly attentive communication style is exemplified by feedback that is direct and explicit. This type of feedback immediately communicates that the analyst has understood what the performer said, or that the analyst requires more information to clarify a potential misunderstanding. Direct, implicit feedback includes nonverbal messages that are responsive to the communication taking place; however, the success of the communication is only implied. Thus, performers can infer that the message has been received based on subsequent questions and nonverbal messages. However, they cannot be sure the analyst has completely understood the communication.

Indirect, explicit feedback is less desirable because, although it is immediate, it may be ambiguous. The performer may not be able to understand the feedback provided or may even fail to recognize that the analyst has provided feedback. Finally, analysts should avoid using indirect, implicit feedback. This feedback does not allow the performer to conclude whether the analyst is paying attention, nor does it signal that the analyst understands the communication.

Questioning Skills. The ability to ask effective questions is useful regardless of the technique. However, some techniques, such as the interview, task analysis, and focus groups require extensive use of questioning skills. Questioning skills include recognizing when a question would be useful in stimulating ideas, ensuring that questions are framed in an appropriate manner, asking a variety of types of questions (e.g., open and closed, primary and secondary), and providing response time in between questions. Chapter 7 discusses effective questioning in more detail.

Listening Skills. Effective listening plays a major role in enabling analysts to elicit the information they require from sessions conducted using any technique. Listening skills encompass attending skills, nonreflective and reflective skills, nonverbal communication skill, and memory (Bolton, 1979). These skill sets, and attributes that exemplify them, are shown in Table 4.5 and explained in greater detail in chapter 7.

TABLE 4.5
Listening Skill Sets and Attributes (Adapted from R. Bolton, 1979)

Listening Subskill	*Attributes of the Skills*
Attending	Minimizing distractions Exhibiting attentiveness Establishing eye contact
Nonreflective	Using attentive silence Using conversation starters Using minimal responses
Reflective	Clarifying understanding Paraphrasing responses Reflecting feelings Summarizing ideas
Nonverbal	Interpreting facial expressions Interpreting gaze and eye contact Interpreting posture and gestures Interpreting personal space messages
Memory	Concentration Short and long term memory

Analyst as Translator and Organizer

Finally, analysts may vary in their ability to analyze session transcripts or tapes, synthesize content, and reconstruct the information into the selected representation or output form. Techniques vary in the amount of effort the analyst must expend to translate and organize the resulting session content. For example, an analyst can easily spend 2–3 hours listening to, transcribing, translating, and organizing data from a 1-hour interview, whereas data from an interactive observation session may require only half that time due to the worksheet used to capture data during observations. Carefully assign sessions and techniques to analysts who not only can deliver the technique and manage it during the session, but can translate, organize, and represent session data in formats the team can use effectively.

Chapter 5

Scenario Elicitation, Analysis, and Generation

Scenarios are example "stories" of normal events and critical incidents that represent the types of situations with which performers must work and use the system (McGraw, 1994e). A scenario has a "storyline," theme, and goals. Scenarios are used to help manage the scope of a complex requirements elicitation activity. This includes narrowing the focus to address the most important aspects of a domain, and communicating emerging requirements among the project team.

Scenario generation, analysis, and development are the primary tools in the Scenario-based Engineering Process. Elicited scenarios are decomposed, using work process and task analysis, to identify the major functions, processes, or "scenes" involved, and the activities they encompass. Analysis techniques are used to understand and develop detailed representations of activities within each scenario.

New scenarios (encompassing required activities and tasks), new ideas for reengineering business processes, and automation opportunities are generated, reviewed, and refined. Prototype screens are developed to portray the initial understanding of system functionality and requirements to users, and to provide a stimulus for the enhancement or refinement of these requirements.

First, this chapter defines three scenario-related activities that drive SEP and form the core domain analysis. Next, it presents guidelines for eliciting and decomposing "as is" scenarios to understand the processes, tasks, performers, information, decision making, and problems involved. Finally, the chapter discusses the development, review, and refinement of "to be" scenarios that define initial requirements and prototypes for the new system, process, or opportunity for automation.

The purpose of this chapter is to enable the analyst to understand what scenario generation, analysis, and development entails, how it relates to other requirements elicitation and analysis techniques, and how it can be used. Specifically, the chapter is designed to help the analyst achieve the following goals:

- Present descriptions of, and rationale for, scenario techniques to customers.
- Elicit and develop current "as is" scenarios using interactive observations.
- Decompose and represent scenarios using a number of different methods.
- Use elicited scenarios as input to work process analysis, task analysis, concept analysis, and decision process tracing.
- Develop innovative "to be" scenarios that meet performer needs and system goals, and that can be used for future validation.

SCENARIO GENERATION, ANALYSIS, AND DEVELOPMENT: AN OVERVIEW

Scenario generation, analysis, and development is the linchpin of SEP. Figure 5.1 illustrates its role in domain analysis and its relationship with other techniques. The three sections that follow provide a framework for understanding how each of these activities helps project teams define requirements and system specifications.

Getting Started: Scenario Generation

Scenario generation involves eliciting and compiling normal and atypical incidents or events (the "as is" scenario). As the first panel in Fig. 5.1 illustrates, scenarios are elicited from expert and novice performers during interactive observation and structured interview sessions to ensure that important aspects of the domain are revealed and understood (Wang, Hufnagel, Hsia, & Yang, 1992). This process involves gathering and documenting the following types of information (McGraw, 1994e):

- Context, including physical context (i.e., topology) and logical context or situations.
- Business circumstances.

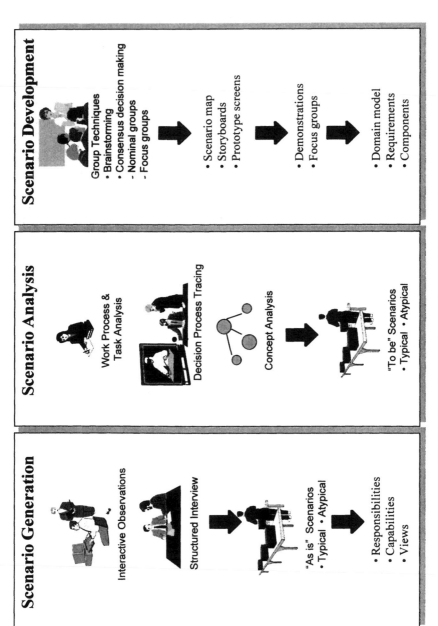

FIG. 5.1. Scenario generation, analysis, and development.

122

- Input data and information used.
- Constraints considered and rules used.
- Specific actions or activities performed.
- Performers and other participants involved.
- Results or outcomes.

Scenarios are elicited and compiled until no new functions, activities, or performers are identified, or until the team has acquired scenarios that reflect each stated desired function of the system.

The elicited scenarios may be documented in a number of ways to assist in the communication of domain experts and performers with the project development team. A given scenario can be represented as a narrative "story" that traces an event to its conclusion. Key scenarios are selected and represented as a three-dimensional structure comprised of responsibilities, which in turn are comprised of capabilities and views. Figure 5.2 illustrates this concept. Responsibilities become the basic element of the reference architecture (see chapter 1). Capabilities include constraints, elements, states, threads, events, structures, entities, and functions. Different views of a scenario include environmental views, views of the domain, and technologically driven views. Together, these provide the project team with a firm foundation for understanding the domain and its requirements.

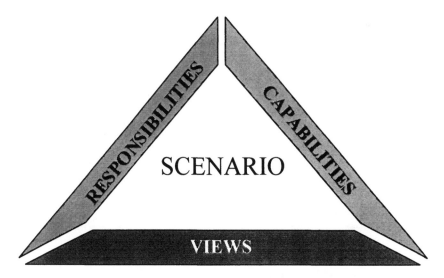

FIG. 5.2. A scenario as a three-dimensional representation of responsibilities, capabilities, and views.

Scenario Analysis: Digging Deeper

Scenario analysis, depicted as the second panel in Fig. 5.1, enables analysts to identify preliminary requirements for an application or set of applications. It reveals interactions among performer groups (i.e., specialists, departments, etc.) and entities within the business environment (i.e., "suppliers"), and explicates and refines roles and responsibilities. For example, by analyzing scenarios, analysts also can identify generic process flows within the system, and specific system transactions among components. Furthermore, scenario analysis clearly reveals time dependent sequencing inherent in the domain activity (Mettala, Harbison, & Hufnagel, 1994) and requirements constraints, values, and determinations.

Scenarios are initially defined at the highest level of abstraction—as major processes or functions. Next, this level is decomposed into tasks and subtasks. The decomposition and analysis of the tasks will result in the determination of initial application requirements. Later, the initial requirements will be refined when the "to be" scenarios are developed.

Using the "as is" scenarios, the project team conducts work process and task analysis (see chapter 8) to identify the following:

- Rationale and goals of the scenario being analyzed.
- Primary processes or functions (often parallel to "scenes") apparent in the scenario.
- Various performers involved in the scenario and the activities (i.e., "roles") for which they are responsible.
- Specific tasks and subtasks performed in completing the processes previously noted.
- Information used and communications required in completing the required tasks.
- Outputs or products of the tasks.
- Specific problems evident in the scenario and a "wish list" of requirements desired in a new application.

Task and subtask descriptions (see chapter 8)—including details about information flow, output, temporal issues, task constraints, and performer roles—provide the input for object-oriented analysis (OOA). Analysis activities, including OOA, result in the development of domain models (Haddock, Kelly, Burnell, & Harbison, 1994) and an understanding of primary concepts and objects in the domain.

After the primary task analysis has been conducted, analysts examine the scenarios and tasks to identify areas in which further investigation is needed. Key targets include those tasks that involve decision making and problem solving (i.e., are not as easy to observe and understand). Further analysis may require that analysts use decision process tracing and protocol analysis (see chapter 10) to define:

- Characteristics of the problem or decision to be made (problem setting issues).
- Features used in recognizing the problem and the appropriate type of solution.
- Hypotheses made, and tests performed to confirm or deny hypotheses during the decision making and problem-solving process.
- Specific steps involved in the decision making.
- Type of mental selection mechanism or rules being applied.
- Alternatives considered, their attributes, and values used in the decision-making process.

Other targets for further analysis may also be identified. For example, identify tasks that are non-value-added, time-consuming, and do not involve performing a function. Tasks that may fall into this category include coordination tasks (e.g., coordinate with, or tell someone something is ready; take something down the hall to someone else). Once identified, analysis can determine the extent to which these tasks may benefit from automation. Also look for tasks that are performed in an inefficient manner or that take inordinate amounts of time or talent to complete. For example, a process may be a vestige of years of working in a certain way, even though that way is no longer efficient. Or, whereas the task itself may require little time to complete, time lags between the steps (while something sits on a desk, is lost and must be recreated, etc.) make it appear to take a long time. These types of tasks would benefit from the application of a new process (e.g., new way of doing the task).

Scenario Development: Envisioning New Systems and Specifications

At this point the scenarios have been dissected and analyzed, and the primary application requirements should be well understood. Roles and interactions have been documented. Performance models, which will be used later in validating the system, have been developed. Requirements constraints and values have been determined.

It is appropriate that the preliminary requirements and specifications, in the form of "to be" scenarios, be communicated among the project team, with performers, and with the customer's management. These new scenarios depict how a new system will function and how users will interact with the system, revealing requirements for functionality, usability, and performance. The new scenarios will stress how the new system or application meets stated goals for the project (McGraw, 1994e). Even if the project does not involve building a new system, "to be" scenarios can be used to define new workplace operator positions or processes.

The new scenarios represent an *initial* specification for the system, process, or position. These scenarios can be used in many ways. We suggest using them to develop storyboards or prototype screens that can help users or customers envision the new system. After users or customers react to and help refine the storyboards or prototype screens, these are developed into more realistic functional prototypes of system components or demonstrations (i.e., "proofs of concept").

Prototype screens, components, and demonstrations are a simplified and incomplete picture of the entire domain, but are intended to address an important set of situations in the problem domain. Once constructed, they can be verified and validated by user groups to refine the initial specification of the prototype system, process, or position description. This refinement process clarifies the omissions and errors in the requirements, domain model, components, and scenarios (Harbison, Burnell, Kelly, & Silva, 1995). Demonstration screens, systems, and prototype components can also be used to communicate possibilities and plans, and to gain management support (especially critical for large projects).

The sections that follow provide guidelines for conducting scenario generation, analysis, and development.

CONDUCTING SCENARIO ELICITATION

Eliciting scenarios is not difficult because humans normally can communicate episodic information easily, as Fig. 5.3 implies. Scenarios are elicited in tandem with other techniques. For example, during an interactive observation session the analyst observes firsthand a scenario that represents a particular type of event. In addition, we often use scenarios to help refocus wandering interviewees—asking them to give an example of a particular incident that illustrates a type of reasoning or a certain type of problem, for example.

It is difficult to determine a priori exactly how many scenarios are enough to depict the range and complexity of the domain under investigation. The following guidelines should help the project team ensure adequate coverage:

1. During project planning meetings, ensure that the customer and the project team agree to the overall functional goals for the new system, application, process, etc.

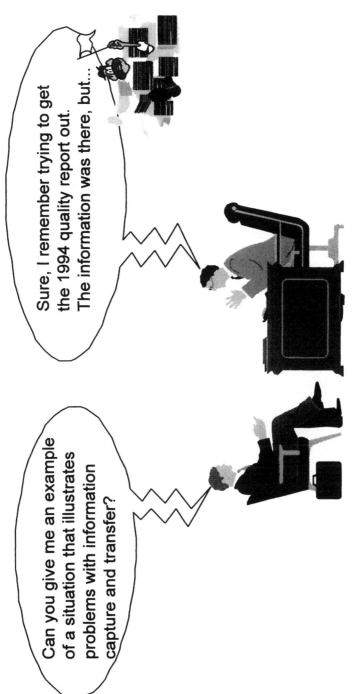

FIG. 5.3. Performers often find it easy to communicate scenarios.

2. Work with the customer to partition the domain into its major functional areas, processes, or phases of activity.

3. Conduct interactive observation and initial interview sessions for all target functional areas. This baseline requirements elicitation activity often reveals other activities, performers, or phases that should be investigated, or special circumstances that impact the basic process.

4. Compile all the scenarios observed or identified during the sessions and categorize these according to functional areas or special characteristics. We recommend that you have enough typical and complex scenarios for each functional area to provide adequate coverage and to identify the common threads and different approaches used.[1]

5. Customer representatives should review the scenarios to help determine any missing areas. These topics or areas are either added to the baseline sessions that must be conducted to elicit adequate scenarios, or compiled as topics for special sessions to explore exceptions, complications, or complexities.

Components of a Scenario

Each scenario captured should be documented in detail to represent the following:

- Goals and critical success factors.
- Physical (i.e., topology, layout) or logical (i.e., business circumstances) context.
- Major events or activities that comprise the scenario.
- Performers or participants involved and the events in which they are involved.
- Information and resources used, including information, products, etc. throughout the scenario.
- Points at which decisions are made, constraints considered, and rules applied.
- Performance problems and opportunities for enhancement.

Goals and Critical Success Factors

Scenario goals and critical success factors provide insight to the viewpoint and level of abstraction of a scenario. The goal of a scenario should be apparent when the analysts asks the question, "What are the performers in

---

[1]"Adequacy" must be judged based on the complexity of the domain, the project goals, available funding, and schedule.

this scenario trying to accomplish?" Events observed are occurring because the performer, company, or enterprise wishes to meet particular goals. For example, trauma care scenarios would share a global goal of "getting the patient the appropriate emergency intervention, transportation, and care as quickly as possible to save life and limb." A specific trauma scenario would share this overriding goal, but may have other more specific goals that drive it, such as "manage emergency medical resources to ensure the successful handling of a multi-victim incident," "avoid over-triaging patients who do not need transport to a trauma center," or "select and apply the appropriate protocol for pediatric patient care, based on the mechanism of injury and primary survey."

Stating the goal for each scenario that has been elicited often happens after the fact, through discussions with the performers involved in a particular scenario. After the scenario has been captured, for example, the analyst can ask performers "What was this all about?" and "What was the overriding goal in the scenario?"

After the goal is stated, the analyst can pursue understanding the critical success factors that help judge whether the goal was successfully met. Critical success factors often include measurement data which can supply initial performance requirements. For example, the following are examples of critical success factors for trauma care scenarios:

- Appropriate units, based on call type, must be dispatched in response to a call within 1 minute of notification.
- Trauma patients should reach the appropriate trauma center within the "golden hour" after their emergency incident occurs.
- Units must be able to reach any address, even remote locations and new subdivisions, quickly (within 3–5 minutes in most cases).
- Medics must be able to assess the patient and determine patient priority within seconds, then quickly begin any emergency interventions.
- Medics must determine appropriate transport and receiving facilities based on patient priority and protocol.

Context

Context includes both the physical and logical contexts for a scenario. The physical context often describes the physical conditions that impact the scenario. This would include the topology of an event—where performers and other roles described in the scenario are located, where activities occur, and so on. Topologies for a scenario can be represented in the form of a scenario map that shows a layout for the scenario (see "Using a scenario map" later in this chapter). Physical context can also be represented as a graphic that illustrates how performers, devices, etc. in a scenario interact with each other.

Logical contexts include conditions, business circumstances, extenuating circumstances, and so on that are not of the physical world, but provide a backdrop against which the scenario events can be better understood.

Major Events

The major events are the primary processes, functions, activities, or phases that occur during the scenario. They provide a high-level abstraction (i.e., concepts and basic information) for the domain and form a skeletal foundation for scenarios in that domain. For example, Table 5.1 illustrates sample high-level events for the primary processes or functions in a scenario.

If the events have a strong temporal or spatial component, this must be documented as well, to provide information that will help ensure adequate performance of the application. Task analysis activities provide a mechanism to decompose these events and examine requirements associated with each of them.

Performers or Participants Involved

Next, the analyst documents who is involved in each of the high-level events. In our example above, we would see medics of various levels of certification, victims or patients of different priorities, and eventually, hospital staff and other support personnel.

Information and Resources Used

Further examination of existing "as is" scenarios should reveal the information and other resources required during each process and event and how they are accessed, used, and manipulated to produce output. Retrieving examples of screen print outs, sample work artifacts, forms used, etc. during interactive observation sessions can help reveal information use. Determining systems and other material used to acquire the information helps analysts understand current means of accessing that information

TABLE 5.1
Partial Processes and High-Level Events for a Trauma Scenario

Process/Function	High-Level Events
Arrival at Accident/Patient	Scene assessment
	Initial assessment
	Rescue requirements determined
Triage	Priority assignment
	Transport and intervention planning
Resuscitation	Protocol selection
	Treatment and intervention

and associated problems. Documenting information use also reveals redundancies in the entry or processing of that information.

Analysts should detail the following:

- Information used, by phase or events.
- Required format of the information.
- Source (system, person, process etc.) and destination of the information.
- Data that comprise the information and the forms, screens, etc. that depict it.
- Reliability and validity of the information.
- Manipulation of the information.
- Problems with the information.
- Security and other factors related to the information.

Decision Points

Analysts also should identify where in the scenario performers make decisions, how these decisions are made, and their importance (see chapter 10). Investigating decision making in scenarios is critical, as this often reveals opportunities for new system requirements that can enhance the process. This makes the process less dependent on an expert performer. Additionally, if the need for required system support can be identified, the system can help performers with less expertise make better decisions. When the analyst identifies a decision point, he or she should document:

- Information and other resources used to make the decision.
- Major constraints involved.
- Alternatives and attributes considered.
- Problems related to making the decision.

Performance Problems and Opportunities

Finally, the analyst reviews the elicited scenarios and highlights performance problems revealed therein. From these the analyst can begin to understand opportunities for enhancements, reengineering, and system support that can reduce performance problems and increase success in the scenarios.

Figure 5.4 is an example of one type of report from a database in which we store information from interactive observations, interviews, and scenario sessions. This particular report prints the "wish lists" from every performer with whom we have worked, coded by session number. This list specifies

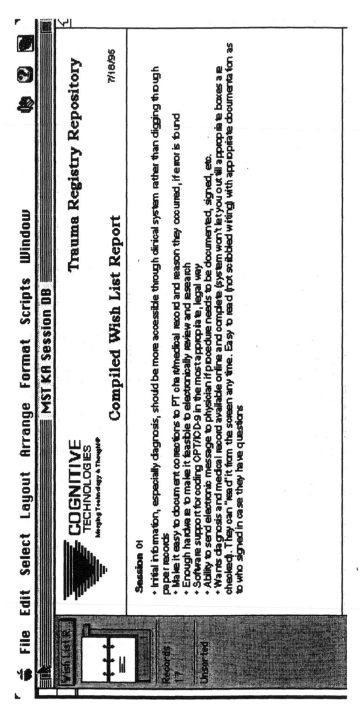

FIG. 5.4. Report summarizing "wish lists" for a new application to reduce performance problems.

what a performer would like to see in the application to reduce specific problems with the process. We can compare this wish list to the "to be" scenario to ensure that we have addressed as many of them as is appropriate.

Guidelines for Scenario Elicitation

The following guidelines will help the analyst capture and document scenarios for later analysis.

1. When planning preliminary site visits and interactive observation sessions, create worksheet forms that will provide support for the documentation of scenarios. For example, a worksheet form that provides the following would be useful (see chapter 6):

   • an area for classification of the scenario according to specific types;
   • a table in which to document major activities or phases, and the screens, work artifacts, and tools used;
   • a boxed area in which to detail aspects of the storyline;
   • performance problems observed.

2. All interactive observation sessions are tape recorded, if possible, for later identification, review, and analysis of scenarios.

3. After a set of interactive observation sessions has been conducted, analysts working on the project (who may have observed in different locations) meet to share scenarios. Scenarios are grouped according to types of events.

4. Different analysts may have used slightly different terms to record the names of major phases or events. They should work with each other and with performer representatives to standardize a set of terms acceptable to all.

5. After grouping the scenarios, analysts review each example in the group to identify the overriding goal and critical success factors for the scenarios in that group.

6. For each scenario group, analysts and performer representatives should agree on the "wish list" of things that would improve the performer's ability to complete these types of scenarios. For example, they would attempt to identify what would:

   • Reduce errors or increase performance accuracy.
   • Reduce time required to complete an activity.
   • Relieve human workload.
   • Increase customer satisfaction.

- Reduce cost.
- Streamline operation.
- Reduce upfront training.
- Reduce negative impact of poor performance (i.e., lawsuits).

7. Summaries of scenarios may be recorded in a database for subsequent review and analysis.

CONDUCTING SCENARIO ANALYSIS

Scenario analysis involves decomposing "as is" scenarios and draft models of how the current system or process functions, for the purpose of better understanding the problem domain. Additionally, we build models that represent the team's understanding of initial specifications for the application system.

Decomposing Primary Events and Activities

Examination of a scenario should reveal the primary events and the activities it entails. This top-level view helps provide a context for the scenario. It increases the team's understanding of the entities and objects involved, the performers and their roles and responsibilities, the activities and tasks, and the flow of information and decision through the system. Useful ways to document the analysis of scenarios include event process traces, graphics, scenario matrices, and models.

Using Event Process Traces

Scenarios with a strong temporal character may benefit from the development of event process traces, which "trace" the activity of each phase and character in the scenario and plot that trace against a timeline. This technique is especially valuable in revealing concurrent processing and activity. It also makes apparent the characters or performers who interact throughout the process. Figure 5.5 is an example of an event process trace that depicts the roles and responsibilities in a scenario and the activities they perform.

Using Graphics and Maps

Simple graphics may also be used to document a scenario, such as a set of graphics that depict the scenario, along with abbreviated text that describes the major activities involved. Figure 5.6 illustrates a simple graphic that conveys the primary activities in a rescue scenario.

Scenario maps provide a topological context for a scenario and are less likely to be used to capture existing scenarios. However, for scenarios that

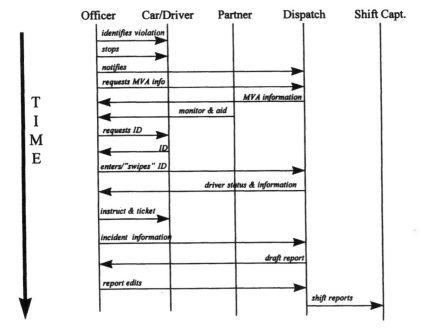

FIG. 5.5. Sample event trace.

are especially complex, or involve numerous activity points, project teams may choose to use a scenario map. (Scenario maps are discussed further in the section "Creating 'to be' scenarios.")

Using Scenario Matrices

Scenario matrices are another tool analysts use to help organize the events, activities, information requirements, performers, and key issues for consideration. A sample scenario matrix is shown in Fig. 5.7. Analysts can easily modify the matrix to meet specific project needs. For example, if communications is a primary requirement or impacts on the successful completion of the events depicted in a scenario, it can be represented as a column in the matrix.

Analysts and designers use completed matrices as a framework for the development of storyboards, prototype screens or systems, walkthroughs, or demonstrations.

Building Models

Models are developed to represent components, services, and the roles and responsibilities of participants involved in the events or processes within a scenario. Initially, these models are humanistic (Hufnagel, Har-

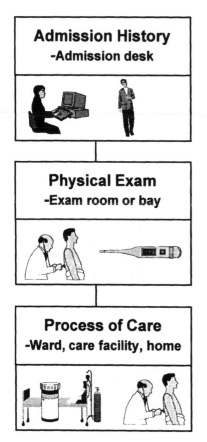

FIG. 5.6. Simple graphics can be used to convey activity and capture a scenario.

bison, Doller, Silva, & Mettala, 1994), reflecting how people view the domain. Analysts refine initial models to depict the objects, attributes, services, and performer roles at a more abstract level. Models are used to ensure that the current domain has been analyzed and documented appropriately, and to provide a structure against which "to be" scenarios eventually are evaluated. Figure 5.8 is an example of a model developed to illustrate the process of a consumer requesting treatment by checking into a hospital (Hufnagel, Harbison, Doller, Silva, & Mettala, 1994). In this model the following activities are illustrated:

1. Patient (consumer) checks into a hospital.
2. Patient registers with receptionist.
3. Patient is examined in the medical department.
4. Procedure is performed on the patient.
5. Medication is administered to the patient.

Scenario	Event	Activities	Information Required	Performers	Key Issues
Traffic violation	Preliminary identification	Identify plate	Plate number	Officer Partner	Quick input of plate ID
		Input plate ID	Data relative to car matching plates	Dispatcher	Can't predict fast response from system
		Transmit plate ID to HQ system			Requires interaction with human dispatcher
		Info relative to vehicle presented on screen			
	Approach				
	Driver ID				

FIG. 5.7. Sample scenario matrix for organizing scenario revelations.

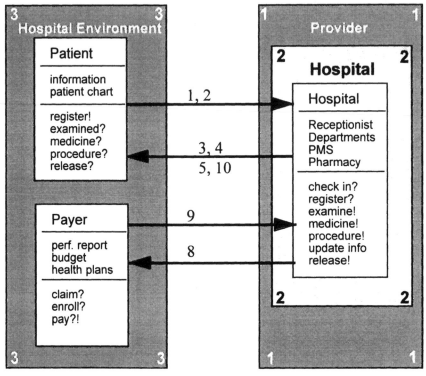

FIG. 5.8. Sample model for a scenario in which a patient requests hospital treatment.

6. Patient's medical record is updated.
7. Patient signs the exit procedure forms.
8. Hospital sends claim for patient's expenses to the Health Alliance (payer).
9. Hospital receives payment.
10. Patient leaves hospital.

CREATING "TO BE" SCENARIOS

Analysts use the scenarios and resulting models to compose the "to be" system into its major components and subsystems. "To be" scenarios reflect the way the *new* system or process should work or be used (including functions, interactions with performer positions, etc.), in typical and atypical events.

Important aspects of the "to be" system or process become more apparent when the project team develops scenarios that describe how the system will or could behave during typical or normal executions. As a result, requirements for normal system operation or process application can be illustrated and communicated more easily. Analysts also generate or develop atypical scenarios, which help identify and communicate error and failure safeguard requirements (Haddock, Kelly, Burnell, & Harbison, 1994).

To create "to be" scenarios, convene a task force comprised of 5–7 people representing both the customer organization and the project team. Representatives from the customer's group should include key performers for the different functional areas that will be reflected in the new scenarios. Other representatives should include managers or supervisors, and if possible, any end customers—the people who benefit from the process. For example, the development of a patient information management system might ask a patient to sit in on the scenario development session, in addition to the more obvious participants—doctors, nurses, admitting clerks, residents, etc. Describe to the group what scenarios are and outline how they will be used to drive system requirements and design. Working with the group, determine how many scenarios should be developed based on project size, funding, and other constraints. Then conduct the major activities described in the sections that follow.

Identify Scenario Goals

Working with the task force described previously, define the goals that the scenarios must help the project to reach. For example, one goal might be to demonstrate the benefits (e.g., reducing errors, shorter time to produce documentation, etc.) of a voice-activated handheld unit for use by telephone repair professionals. This goal can be realized through many story lines. One resulting story line or scenario might feature use of a hands-free device to document findings in real time, and another might demonstrate the impact that real-time capture of data has on the quality of information about a troubleshooting call. Several scenarios may be required, depending on the "view" of the system on which attention will be focused during the demonstration or prototype presentation of the scenarios. (Analysts may need to use techniques such as consensus decision making to facilitate the team's definition of, and agreement on, the goals.)

After the main goals of the scenario(s) have been determined, the task force continues to work together to identify the characteristics of each desired scenario (in general, what it should include, what types of problems it should address, etc.). Then they outline the high-level activities that should take place and the key features each scenario will encompass.

Document and Represent the Scenarios

After this initial definition of the scenarios is complete the project team documents and represents the scenarios using scenario maps, flowcharts, event process traces, and matrices. The primary purpose of this activity is to represent the "to be" scenarios in ways that can be communicated easily to others, enabling the identification and refinement of requirements for functionality, information flow, interactions, and performance. The following sections describe different techniques for communicating scenario-related descriptions. The type of domain being modeled should determine the selection of the appropriate representation.

Using a Scenario Map

The easiest representation, and the one with which we usually begin, is the topological map. If the domain represents activity that occurs in different geographic locations, is handled by multiple performers, and occurs over a long period of time, a topological map will be required to help frame the context. Figure 5.9 is an example of a topological map that helps define the context in which the scenario will occur. This example depicts an environment in which a public safety management information system might operate. At each location in the field, a different mini-scenario will occur (e.g., robbery, mugging, officer down, motor vehicle crash). The system to be used by the officer at each site will enable the communication of information between him or her, headquarters, and 911 dispatch.

Topology maps or drawings are also useful to document areas inside buildings and work places. However, if the domain to be modeled represents activity that occurs in one place, by one performer position, within a confined period of time (i.e., at individual workstations within a call center), a topological map may not be necessary.

Using Scenario Flowcharts

Informal scenario flowcharts are used to depict the major scenes or events for a scenario. Together, the boxes within a flowchart comprise a skeletal script for the scenario. Each box is labeled to indicate the main activity within a scene; text beneath each box provides further detail about the activity. Figure 5.10 illustrates a partial scenario flowchart for the second scenario in Fig. 5.9.

Creating Scenario Flowcharts. A scenario flowchart should be created for each scenario being considered. Creating a scenario flowchart is best done by bringing together the group of analysts who did the interactive observation and focus group sessions with the performers. They will have the

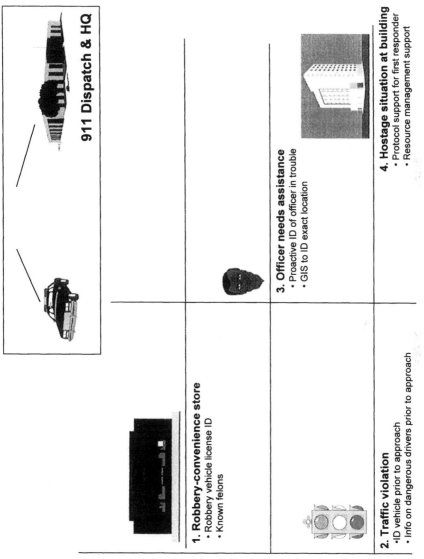

911 Dispatch & HQ

1. Robbery-convenience store
• Robbery vehicle license ID
• Known felons

2. Traffic violation
•ID vehicle prior to approach
• Info on dangerous drivers prior to approach

3. Officer needs assistance
• Proactive ID of officer in trouble
• GIS to ID exact location

4. Hostage situation at building
• Protocol support for first responder
• Resource management support

FIG. 5.9. Topological map to frame the scenario context and communicate events.

Traffic violation scenario:

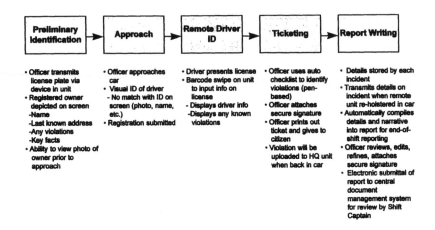

FIG. 5.10. Sample scenario flowchart depicting major events and activities.

best understanding of current scenarios and how to weave together the new requirements and functionality the performers desire. It may also be useful to ask a representative from the customer's organization, such as a supervisor, a performer, or project manager, to participate in these sessions. This ensures that the scenarios reflect the needs of the organization, that their feasibility has been considered early, and that questions that arise about the domain or process are answered quickly.

Using techniques such as brainstorming and consensus decision making (see chapter 11), one analyst facilitates a session in which the participants work together to expand each of the scenarios. At this point the level of detail in each box of the flowchart must depict the primary events or activities, but does *not* have to be extensive. The text beneath each box must include information about the high-level tasks the performer would accomplish. Later, this information will be used to develop sets of storyboards and sample screens to support each activity box. For example, to accomplish the activity or event in the first box, Preliminary Identification, the performer might have to view and/or interact with two or three screens in the system.

Selecting Scenarios to Pursue. It is often the case that the team can identify more useful scenarios than it is possible to develop, given budget and schedule constraints. In this case the task force should meet to reach consensus as to the scenarios that should be developed into a working prototype or demonstration system.[2] To select scenarios to develop into

---

[2]Scenarios not selected for development should be kept for use in defining requirements and functionality of components within the system.

prototypes or demonstrations, analysts can use a group technique such as the one that follows.

1. Convene a meeting of 5–7 stakeholders in the system. These individuals should include a representative from project management, two key performers, one or two supervisory personnel who understand the impact of performance improvement, and analysts who participated in the initial requirements elicitation activities.

2. Prior to the meeting, develop a scenario flowchart representing each of the potential scenarios, and post these around the room. If the flowchart is filled with text and boxes, we recommend attaching the flowchart sheet(s) to a posterboard or section of Kraft paper prior to posting them on the wall. This will leave room for participant comments.

3. On their arrival, participants are given a packet with Post-It™ style sticky notes and sheets of three different colors of sticky dots or round labels. (We use no larger than ½″ diameter round labels.)

4. First, conduct a general presentation that provides the participants with a background on how the scenarios posted around the room were selected. State that the goals of the session are to (a) select the scenarios that illustrate the "hottest" or most desired functionality and (b) refine these selected scenarios.

5. Present each scenario to the participants, pointing out why it was developed, its key features, the major activities involved, and the type of interaction a performer would have with the system. The facilitator may use a scenario map or flowchart and should take no more than 5 minutes to present each scenario.

6. Lead a discussion in which members brainstorm to identify the major pros and cons of *each* scenario. Ask participants what they like best about each one, and what disadvantages they see. This discussion should be conducted at a general level, and should not take more than 20–25 minutes.

7. Ask participants to use the round colored labels in their packet to help identify the scenarios that provide what they feel are the best opportunities for prototype screens and demonstrations. They will place a colored dot on each scenario to indicate how they would classify it. For example, if it is "hot" compared to the other scenarios and they believe it is critical that the scenario be developed into prototype screens, they would attach a red dot label. If it is unimportant, unimpressive, or could be distracting to the purpose of the project, they would attach a blue label to it. If the scenario is good, and would further the purpose of the project, but is not extremely exciting, they would attach a green label.

8. Give participants approximately 10–15 minutes to walk around and label the scenarios with their dots. Then have the facilitator lead a discus-

sion that allows participants to express why they did or did not want to include a scenario.

9. At the end of the discussion, the facilitator should identify the scenarios with the most "votes" using a weighting scheme. Red labels can count as two votes, green labels can count as one vote, and no votes are allotted for blue labels. Voting should reveal a consensus as to which scenarios should be selected.

10. Have the facilitator remove the posters for scenarios not selected (but keep them in a file to use when determining requirements). The facilitator then asks participants to take the yellow sticky notes and write one suggestion for refinement (or other editorial comment) per sticky note and to attach as many notes as needed to a selected scenario. This provides participants with an opportunity to refine the selected scenarios, identifying additional activities or details that will be needed. The refinement phase should take no more than 20–30 minutes.

11. The facilitator summarizes accomplishments and thanks participants. After the session is over, he or she compiles the notes for each scenario and begins the refinement process. Results should be communicated with each participant.

Create Prototype Screens and Components

Prototypes bring the requirements and functionality embedded in a scenario to life. Prototypes are constructed to the degree of complexity necessary to demonstrate that the team has adequately addressed features and problems of the domain. This is determined by deciding the depth or extent of the functionality that will be prototyped, and the breadth of the features to be included in the prototype (Asur & Hufnagel, 1993). The project team can use one or more accepted prototyping approaches, including visual and functional prototypes.

A visual prototype is a representation of what the system (or process, application, or new position) might *look* like. It provides a vision of what the system might do and depicts realistic examples of actual screens, bringing scenarios and storyboards to light. However, it is not a functional prototype, nor is it complete in depicting all the system could do. Visual prototypes are useful in presenting hypothetical applications to users. The primary purpose of a visual prototype is to illustrate functional and interface requirements, as they are currently understood, and elicit feedback from performers and customers. Prototyped screens help users and performers respond to and inspect an interactive scenario. Feedback from them helps identify areas in which requirements must be changed or clarified. This leads to the identification of suggested refinements to the requirements and specifications illustrated in the prototype.

Functional prototypes provide more depth of functionality. They enable analysts to give performers and other reviewers a good feel for how the system actually could operate. For example, a functional prototype would enable analysts to demonstrate a complete set of selected functions and screen operation. Furthermore, developers can set up a test bed to test not just the accuracy of how they have visualized system functionality, but also performance issues and feasibility of initial requirements.

Regardless of the type of prototype used, it is an incremental or evolutionary process. Prototype screens, components, and systems evolve continuously. Eventually, these prototypes can evolve into a complete system or can be discarded after the definition of components and requirements.

Using Storyboards

One of the tools we use to manage the development of prototypes for our scenarios is storyboarding. Storyboards are drafts of preliminary screens, or interactive screen displays, that help illustrate how system screens might look as a scenario is being carried out. Storyboards illustrate the types of screens with which the performer might interact while using the new system to respond to the events in a scenario (McGraw, 1994e). Storyboards also include text that details what actions the performer takes when interacting with the screen.

Storyboards can be printed and used to stimulate user feedback or edits. Alternately, performers can watch a computer animation of the storyboards, which simulates how system screens look and change as the user works through the scenario.

The goal of the storyboard development and review process is to present a representation of the flow of screens that can be refined prior to prototype development. Storyboards also may be used to depict and reengineer a workplace or scenario in cases in which a new system is not the end goal.

Storyboards are reviewed with performers (McGraw, 1994e; Hsia & Yuang, 1988) to accomplish the following:

- Help ensure that screen design elements correspond to the performers' mental models.
- Evaluate whether reengineered functions, screens, or reengineered work places actually meet performer needs.
- Reflect the analyst's view of the human-computer interface requirements so they can be clarified.
- Obtain immediate feedback that can be used to refine the developing interface.
- Validate and communicate the real intent of the requirements.

After performers review storyboards that support scenarios, the analyst reports on the results of the evaluation and makes design changes. The sections that follow summarize guidelines for the development and refinement of scenarios.

Guidelines for Developing "To Be" Scenarios

1. Work from a firm understanding of the domain. This may require compiling models and object diagrams to illustrate performer roles and responsibilities, objects, and activities.

2. Compile a list of the desired functionality communicated by performers, and requirements determined by analysts based on interactive observations. Analysts are cautioned to manage the process of eliciting performer desires and wish lists. There is a tendency to develop a shopping list of features into a scenario or prototype. This may not be feasible in the actual system and could lead to disappointment in the resulting application (Carey, 1990). If a reality check is needed, developers can perform trade-off studies on important but questionable wish list items.

3. Using the scenarios elicited during interactive observation sessions, completed task analyses, and lists of current problems and desired future functionality, define a series of potential scenarios. This definition should include the name of the scenario, its goals, and the critical success factors it should demonstrate. These potential scenarios can be depicted in a matrix that allows comparison and may lead to a combination of a number of scenarios. This matrix should be reviewed and refined by analysts familiar with the domain.

4. Using the refined scenario matrix as input, create a top-level scenario map or flowchart to illustrate the "big picture" and interaction among multiple scenarios (in the case of a scenario map), or the major events and activities involved in a scenario (in the case of a flowchart).

5. Convene a meeting with selected performers to select scenarios for development into storyboards and prototypes. This team compares the strengths and weaknesses of potential scenarios and selects those that have the best power to communicate desired functionality in a manner that appears feasible.

6. For each selected scenario, determine the type of prototype to be developed, or if a combination of prototypes will be used. For example, the team may decide that four scenarios should be prototyped, but only one scenario will be developed into a functional prototype, illustrating functional depth in a particular area, but limited breadth. The other three will be visual prototypes with less depth and more breadth.

7. For each scenario, develop a storyline that describes how performers interact with the system to complete the tasks and activities defined in the scenario. We have found it useful to assign one or two people the task of working with analysts to flesh out the storyline in each scenario. This

provides more consistency than if the teams worked separately. The storyline for each scenario should be documented. Many possible formats for documenting storylines exist, including scenario scripts and outcome matrices. Figure 5.11 illustrates a portion of a script for an extensive set of demonstration scenarios.

8. Assign analyst and designer teams to each scenario to begin the storyboard development process. All teams meet together to develop initial screen specifications, templates, and draft icons to enable teams to work independently later, while still ensuring consistency in the way screens

---

Event: Townhouse/apartment fire
Setting: Central City and surrounding counties. Multiple jurisdictions surround the Central City area.
Features: • Fire takes place in an area which can be serviced by multiple jurisdictions
              • Requires resource planning
              • Scene escalation
              • 1 dead; 1 decoy (cardiac history w/smoke inhalation)
              • Multiple high priority patients

Event initiation:
A neighborhood resident calls in on 911, hysterically reporting a townhouse fire in an area of Anne Arundel County that borders Prince George's County, the downtown area. The dispatcher has to calm the caller to be able to get the details. The following information is conveyed:

        Call in: 3:45 PM
        Estimated time of event: 3:40 PM
        Address: Trafalgar St., between 301 and First St. in Bowie, Prince George's County
        Problem: Fire and smoke
        Known casualties: Unknown
        Caller: M.B. Jones, 356-7865

1.0 Dispatch and Mobilization Activity
Notes: • Responders are warned that scene may be dangerous.
          • First responders conduct scene assessment and note that the scene involves a partial collapse of at least one of the units, and that there are multiple casualties. Victims have already been dragged out; no one else is suspected of being inside. They contact Dispatch and request units with ALS capabilities. Two additional units are dispatched. The total ambulance units responding initially are as follows:
          • ALS unit—1 driver, EMT-P, 1 ALS certified
          • ALS unit—1 driver, EMT-P, 1 ALS certified
          • ALS unit—1 driver, EMT-P, 1 ALS certified
          • ALS unit—1 driver, EMT-P, 1 ALS certified

FIG. 5.11. Portion of a scenario script.

look and are used. Using a selected software application, each team creates screen sketches to support the activities that occur in each event or process in the assigned scenario. Each screen sketch should be accompanied by a text or notes field that provides a description of how the user interacts with the screen. Alternatively, screen sketches can be used to illustrate the storyline developed for the scenario.

Guidelines for Reviewing and Refining "To Be" Scenarios

1. Storyboards are reviewed by a group of performer representatives (i.e., performers, supervisors, etc.). The analyst team presents the group with the scenario map to give the big picture of the storyline they will be seeing in storyboard form. Next, they present the storyboards, describing the events and activities each storyboard supports. After each storyboard is presented, the facilitator solicits comments, asks specific questions about issues that may still be unclear to the analysts, and identifies necessary refinements (see chapter 12).

2. Analysts and designers document all changes to the storyboards and begin the development of the prototype. Prototype development activities and tools will differ based on the prototype desired.

3. Prototypes are demonstrated to a small group of performers for validation and refinement. At the beginning of the session, the definition and purpose of a prototype are described. Participants should understand that prototypes have limited capabilities and capture only the essential features of the application (Asur & Hufnagel, 1993).[3] The scenario should then be presented as a story, with the prototype being used to support how the events transpire and how performers interact with the system in the scenario.

4. Issues raised during the preliminary validation and review of the prototypes are investigated by the project development team and members of any review panels (see chapter 12). This group determines which refinements are important enough to include in the prototype, and which can simply be documented as requirements.

5. Scenarios may be prioritized according to some criteria (e.g., feasibility, priority viewpoint, etc.) and used as the basis for a test plan.

MOVING FORWARD

This chapter described the three-pronged process of scenario generation, analysis, and development. The scenario process helps analysts build a better

---

[3]Note: Analysts should manage the expectations of the participants to avoid later disappointment in the actual system.

understanding of the domain. This process requires the use of elicitation and analysis techniques such as interactive observation and interviews, workplace and task analysis, decision process tracing, and group techniques.

We analyze scenarios to better understand the domain. Doing so requires that analysts use techniques such as task analysis, which reveals tasks, subtasks, decision points, information usage, and related performance problems. Additionally, if the domain requires problem solving or decision making, analysts may be required to use decision process tracing and protocol analysis. Regardless of the type of domain, analysts should perform concept analysis, which helps define the entities and objects (both physical and cognitive) used in completed domain tasks and decisions.

The generation and analysis of "as is" scenarios and the development of new "to be" scenarios requires a considerable amount of interaction with groups of performers, stakeholders, and other reviewers. Analysts should be able to use a number of group management and facilitation techniques to ensure the effectiveness of the groups and the products they create.

Chapter 6

Conducting and Analyzing
Interactive Observation Sessions

Summary descriptions or interviews alone don't provide information about how an activity is really performed and often don't reveal performance-related problems. Only by being present during the activity can we see, and later discuss, the details of what helps and hinders work. Simply talking to someone outside the context of actual work may result in top-level responses devoid of details.

Interactive observation is a technique that enables the analyst to observe and work with performers within the context of their job environment. It entails actual observation of work flow and its artifacts. This is followed by brief questioning segments in which the analyst tries to understand better what is happening, how the performer is approaching the work, and what problems may be present. It is one of a class of techniques that is frequently used to obtain data by directly observing the activity under study (Drury, 1990). Other terms that may be used to describe similar techniques include *contextual inquiry* (Holtzblatt & Jones, 1993), *participant observation*, or *direct visual observation*. It is often used as a precursor to more detailed analyses, such as work process and task analysis.

You might say that interactive observation is just common sense. As logical as it seems, however, it is not occurring on all projects, and when it *is* attempted, it is often misused. We recall a *Wall Street Journal* article a few years ago that described the misadventures of two young knowledge engineers as they sought to develop a knowledge-based system to identify problematic fissures in dams. It was only after several months of interviews, the elicitation of numerous rules, and several thousands of dollars that they went on site with the expert. During these onsite trips they discovered how he really did his troubleshooting, and what he really considered. They learned first hand the difference that soil type, dam material, and other important variables made in the diagnosis. However, the learning was too little, too late—the customer did not fund the follow-on phase.

We relate this story not to degrade the knowledge engineers, but to stress the importance of using interactive observation in the Scenario-based Engineering Process. In fact, you could say that this technique is the foundation of the process itself, in that it is often through these types of sessions that scenarios are identified, generated, categorized, and catalogued for later analysis.

In this chapter we describe the purpose and goals of interactive observation within SEP and provide background on why and how it should be used for systems development. We structure the process, detailing how to get started and what to expect at each step. We describe how to develop a sampling schedule, if it is needed, and how to develop observation templates that assist in recording and analyzing session information. Next, we discuss the structure of a typical interactive observation session. We define your role as observer and questioner, and illustrate the types and levels of questioning that are appropriate. We present samples of output from interactive observation sessions and how it can be analyzed and used. Finally, we highlight the advantages and disadvantages of the technique.

INTERACTIVE OBSERVATION: PURPOSE AND GOALS

Interactive observation is sometimes confused with interviewing. However, its purpose and goals are very different.

The purpose of interactive observation is to stimulate the revelation of the normal work process or task performance within the natural context of the job. It enables us to understand work activity from the performer's perspective. It is appropriate when the work process or tasks being analyzed are amenable to observation—that is, have physical/kinesthetic, visual, or verbal components.

In addition to revealing work within natural settings, this technique can also be used to stimulate the generation of scenarios (see chapter 5).

The goals of interactive observation include:

- Observe the performer in the natural context of work—the job environment.
- Conduct observation in as unobtrusive ways as possible.
- Document actual work activity as opposed to prescribed work activity or protocol.

- Capture significant visual events.
- Capture significant auditory events.
- Capture naturally occurring scenarios (see chapter 5).
- Verify observations through the use of interactive questioning and discussion.
- Capture work-related events and scenarios in a format that makes later analysis easier.

STRUCTURING THE OBSERVATION PROCESS

Unless these sessions are structured and planned carefully, they can easily degenerate into hours of taped recordings that are meaningless or that capture insignificant events. The sections that follow offer practical advice in setting up and conducting well-planned, structured interactive observations.

Getting Started With Interactive Observation

If you accompany performers as they accomplish work activities in their work place, and focus them on their experiences, you can quickly gather usable, concrete information. The "mental anchors" they see act as stimuli to help them remember details and recall specific experiences. The result is often that your observation of them in the work context reveals more usable information than other kinds of information gathering. However, before you schedule individual interactive observation sessions, you should complete the preliminary tasks that follow to ensure the success of the sessions themselves and encourage better participation from performers.

- Take a "quick tour" of the work environment.
- Complete an initial categorization of activities you wish to observe.
- Determine if sampling will be necessary.
- Identify and help select performers to observe.
- Plan the sessions, and communicate with performers.

Taking a Quick Tour

Who among us would visit a foreign country without previously consulting a map, reading about its people and cultures, or learning a basic vocabulary in the accepted language? Planners for military missions spend a great deal of time on these preparation activities prior to going in-country to accomplish the actual mission.

As analysts, we must be similarly prepared. Prior to setting up an observation schedule, selecting performers, and taking up their valuable time, we must determine what to observe, where we will observe, an observation schedule, and how much observation is needed (Whiting & Whiting, 1990). The following steps will help you structure an interactive observation session that meets your project goals.

1. Determine clear goals and the desired outcome for your interactive observation sessions. Defining the type of information you expect and desire from the session can help you determine if interactive observation is the most appropriate technique, and what type of recording (e.g., audio tape, video tape, notes) may be most useful. (See chapter 4 for assistance in selecting the right technique.)

2. Determine how interactive and obtrusive your analysts will be. This can range from unobtrusive video recording followed at a later date by an interview, to sessions in which the analyst observes and actively queries the performer. (See chapter 10 for a discussion of process tracing and protocol analysis.)

3. Determine if the selected means of recording will actually work in the job environment to produce the output required. For example, video recording of monitor screens results in playback that includes screen refresh lines that may make it difficult to see what you had wanted to record. Background noise may make audio tapes unintelligible.

4. Understand the different divisions, departments, or work groups that are involved in the process you wish to observe. Determine how long it takes to interact with another worker or department, and how transfers of information occur. This impacts decisions such as how to record these interactions, and how to document cooperative tasks that are undertaken concurrently by two different performers.

5. Be aware of cultural issues that might impact the success of your interactive observation sessions. These include both societal and corporate culture issues. For example, observation sessions held for a company with multiple call or distribution centers occurred in centers in each region of the United States because the cultural mix of employees differed by region. Corporate culture must also be addressed because different performer echelons exist and intermingle with varying degrees of success at different corporations. If multiple echelons are all involved in the processes to be observed, it is critical that the observation sessions include each performer group.

6. Become familiar with primary concepts, terms and specialized vocabulary, computer systems used, etc., to increase the effectiveness of your questioning activities.

All of these goals can at least be partially attained by going on-site for a quick tour in which you and a few other key members of your team familiarize yourselves with the work environment, its components, and work groups. The amount of time you spend on the quick tour depends on the complexity of the environment and the numbers of performers involved in it.

For example, a three-person team conducted a quick tour of a call center environment in a 3-hour time period. The project, in this case, was confined to a call center environment in which the performers used one primary computer system that linked to other, secondary systems. The performers represented three work groups or specializations: domestic, international, and freight representatives. After completing an overall tour of the center, each of the people on our team rotated through several short (20-minute) observation cycles in which we sat with (and listened in on) performers as they worked with customers. We captured initial information such as:

- Primary work goals and very high-level accomplishments.
- Types of calls each performer type handled.
- Overall ease with which performers interacted with the existing computer system.
- Screen captures for activities that occurred frequently (i.e., took high percentages of their time).
- Vocabulary/terminology used.
- Resources accessed during the short quick tour cycle.
- Scenarios that occurred that were representative of the process.

Later, our group met together and analyzed what we had seen. We created a list of the different types of activities we had observed and grouped these in a logical fashion. We identified primary vocabulary and acronyms, and began a glossary which would later become the foundation for our domain dictionary (see chapter 9). We compiled a list of things the performers had said about work problems, motivational issues, hardest thing about the job, etc. We also compiled sets of scenarios we had seen. From this, we developed a presentation that reviewed the findings of our quick tour activities and suggested a course of action for the individual interactive observation sessions.

The quick tour can be more complicated if the domain is large and complex. For example, another project required that we analyze the complete work process of responding to trauma events. This process involves many different job functions—dispatchers, medics, surgeons, nurses, flight crew, ambulance drivers, patient administrators, etc. It varies depending on whether the domain is military, metropolitan/urban, or rural. The quick tour for the metropolitan phase of this project was termed a *blitz* and involved

2½ days of interactive observation. Small teams (two or three people) spent up to 3 hours in each of five "stations," representing the different job functions involved in the overall process. The stations toured included:

- 911 center, in which dispatchers, supervisors, and computer systems professionals interacted with callers and entered information into a dispatching system.
- Individual ambulance and fire response stations, where paramedics, drivers, and other professionals received the dispatch and responded to the event.
- Area-wide resource management and communication system operations, where dispatchers monitored the status of area hospitals (certified for various levels of trauma care) and medevac helicopters, enabled communication between emergency responders and physicians, and assisted in directing ambulance and helicopter units.
- Hospital emergency rooms with lower-level trauma capabilities and specialty units, where nurses, surgeons, attending physicians, and administrative personnel assist trauma patients.
- Hospital trauma resuscitation units with higher-level trauma capabilities and specialty units, where nurses, surgeons, attending physicians, and administrative personnel assist trauma patients.

Figure 6.1 is an example of a schedule from this blitz activity. It indicates which stations the analyst would visit and the times he or she was expected to tour that station.

Session Time	10/17	Performers
9:00–10:30am	Smithson Co. EMS 1456 Jenson Rd. Lutherville, MD Contact: Lt. J. Smith	Roger Martin Marianne Criz Lt. J. Smith
11:30am–1:00pm	St. Joseph's ER 47 N. Main St. Elliott City, MD Contact: Dr. Brad Jones	Emily Southern, R.N. M. Luke, Admitting Dr. Brad Jones
2:00–3:30pm	Washington Co. Medevac 1114 Long Pkwy. Washington, DC Contact: Ed Bruner	Rue Michaels, flight nurse Lt. Conrad, pilot Ed Bruner, dispatch

FIG. 6.1. Sample quick tour schedule for a complex domain with multiple job functions, performers, and job sites.

At the completion of this blitz of activity, we summarized what we had seen, shared it with other team members, and identified what we needed to target for interactive observation and task analysis sessions. Information gained from this activity was deemed very valuable to the project members. They gained a better understanding (and appreciation) of each performer's role in the total process and began to familiarize themselves with the concepts and vocabulary in the domain. Other benefits were realized, as well. The performers themselves became acquainted with our project and analysts. Later, when we called to set up actual sessions they were less fearful and more eager to cooperate with us.

Categorizing Activities

To conduct an effective interactive observation session and record information in a meaningful fashion, you must know what you are looking for and what you are seeing. After you have observed the process during the quick tour you should have an identifiable list of activities that occurred. Compile this list of activities, categorizing them in a way that makes sense based on the work flow. For example, activities discovered during a quick tour for a regional retail management information system included those shown in Fig. 6.2.

These activities represent the framework for the eventual interactive observation sessions held at regional sites across the country. Keys to effective activity identification and categorization include the following:

Eight Primary Activities Captured During Observation	
Review store profits and losses • Daily report analysis • Weekly report analysis • Monthly report analysis	**Determine cause when retail goals are not met**
Review the effectiveness or impact of marketing campaigns on retail sales	**Identify store managers with retail goal exceptions** • Exceeding their goals • Not meeting their goals
Identify stores with higher than expected losses	**Determine root cause of higher-than-expected losses**
Prepare reports • Accounts receivable and payable • Personnel reports • Inventory reports	**Manage store managers** • Hiring • Firing • Evaluating • Training

FIG. 6.2. Types of activities catalogued during a quick tour session.

- As you identify an activity, determine if it is something you will be able to view and record during an observation session.
- Try to keep your list of activities to 20 or less (Kirwan & Ainsworth, 1992) unless you plan on videotaping. It is very difficult to discern which activity you are viewing, scan and select it, and record observation comments if your list of activities is longer than 20.
- Make sure the categories of activities you identify are distinguishable from one another by your analyst team. You want to ensure the likelihood that two different analysts would recognize the difference between two activities and categorize the observed activity properly.

After the categories have been defined, compile them into a worksheet that analysts can use during their observations. Figure 6.3 is an example

| Performer: _____ |
| Process: |
| ___Dispatch   ___Scheduling   ___Tracking   ___Rating   ___Adjusting |
| Stimulus/cue: _____   Goal:_____ |

Activity	System Screen	Field	Questions Asked of Customer

| Specific problems encountered: |
| Other tools, references used: |
| Work artifacts produced: |

FIG. 6.3. Body of a sample interactive observation worksheet for use in gathering data within the context of the job environment.

of an interactive observation worksheet and supporting questionnaire we used on a recent project. This worksheet was devised to make it easy for the analyst to identify the appropriate activity represented by a customer's call, and to relate performer queries directly to each call type. Other information about the call is captured for use in scenario analysis.

Sampling During Observation

Sampling is a method analysts use to gather information about the time spent on an activity, the goal being to define the *proportion* of time spent on each activity in an observable process.[1] It involves observing the performer at specific intervals and recording the activity in which the performer is involved. It is an objective yet time-consuming method that documents temporal aspects of an activity. It helps you understand the percentage of the job that a particular activity consumes. This information helps you determine the activities consuming the majority of a performer's time, and can provide clues to your system design.

For example, we were building an embedded performance support system into the user interface of a system. Part of our job was to identify performance problems and their impact to enable us to design screens and supporting advisors, coaches, and help that would enhance performance. The customer provided activity sampling data that indicated performers spent approximately half of their time on a single work process—dispatch. They believed that little support was necessary for this work process because it was so common and was not perceived as difficult by supervisors and experts. However, the fact that it comprised nearly 50% of the job, combined with reports of the following factors, led us to suggest enhancements and support:

- Performance problems by novices.
- The amount of training required to reach optimum performance.
- The potential cost of errors (both hard costs and soft costs like customer perception of satisfaction).

Thus, performing sampling or reviewing sampling data is easy. It is what you *do* with the data, and how you interpret it, that determines how useful this method will be. The following sections provide information to help you set up and conduct a session using sampling.

Why Sample? Sampling is useful if you need to understand the relative frequency of observable activities that comprise a process. It also helps you validate information about a job's tasks that may have been provided earlier by a performer or supervisor. It may reveal that some activities are being

---

[1]This process was first described by Christensen (1950).

performed that were not defined as being part of the job (whether they are actually part of the job or whether they are superfluous, detracting activities).

Sampling is unnecessary if relative frequency data is already available and you are confident about its accuracy. If key activities are cognitive and not observable, or if you do not need this type of information to determine functional or usability requirements, do not use sampling.

Defining the Sampling Schedule. Designing the sampling activity will take more time than conducting the actual session! This involves ensuring that observable categories are complete, creating the sampling schedule, creating the worksheet, and gathering the timing devices you need. However, time spent preparing for the activity pays off when you conduct the session and attempt to interpret the data you have gathered.

As with any observation session, you should document the categories or types of activities you want to observe prior to conducting the session. Some researchers (Kirwan & Ainsworth, 1992) suggest a short pilot study to ensure that the categories of tasks selected are complete and observable.

Next, determine how often you need to sample, and define a sampling schedule. This enables you to note scientifically the activity that is occurring at a predetermined, fixed interval of time.[2] Obviously, the analyst should consider the length of the activities to be sampled in defining this schedule. For example, ensure that the sampling interval you set will enable you to observe even the shortest significant activity.

Finally, you must define the total amount of time the sampling will require for any specific observation. Of course, you must understand the approximate length of time the tasks being observed take to complete to ensure that you will have an opportunity to record relative frequencies of tasks within the total target activity.

Creating the Sampling Worksheet. After you have identified the activity categories and the sampling schedule, construct a worksheet to use during the sessions to record the desired information. The simplest worksheet is a tally sheet. As you observe, note what the performer was doing at the predetermined time interval by placing a tally mark beside the category that most closely represents the activity. Figure 6.4 illustrates the simplicity of this tool. Note, however, that the tally method provides frequency data only. It will not help you determine how long a particular activity takes, or necessarily, the sequence in which activities are performed.

---

[2]Although random sampling intervals can be used effectively, especially for activities which are very lengthy, they provide more limited information.

Performer:_____		
Sampling schedule:_____ Total time observed: _____		
Observable activity	Count	Total

FIG. 6.4. Example of a tally sampling worksheet.

Additional Sampling Tips. When combined with interactive observation, sampling provides useful information that helps the analyst understand the work process better. This method is most effective if analysts follow these guidelines:

1. Use sampling only when necessary. Not every project requires that you understand relative frequencies of activities within the total work process. Sampling should be used for observable tasks and should not be attempted for cognitive activities.

2. Ensure that the categories of activities you defined are complete and accurate. Nothing is worse than recognizing during the sampling activity that you have failed to document a category accurately, or have left it out entirely. Spend time up front (on the quick tour perhaps) carefully documenting the activities involved. Just in case, construct the worksheet with extra spaces for those other activities that you may discover.

3. Use a tool that signals you when the next interval has begun. Don't rely on your ability to estimate time (we usually do that poorly), or think you will remember to look at your watch systematically. If the performer's organization will allow it, the best approach is to videotape the observed activities, then pull sampling data from the video.

4. Be careful how you use sampling data. As you interpret the data you collected, you may be inclined to equate more importance to activities that occurred with greatest frequency. Remember, this data only provides information on relative frequency—not relative importance. You may find an activity that is not performed very often but, in fact, is more important (e.g., based on cost of errors, consequences of errors, customer satisfaction, etc.) than a very frequent activity.

Selecting Performers to Observe

The key to the success of an observation is the people you are told to observe! As obvious as it sounds, you must ensure you are observing *performers*. Our experience has shown that a client's first tendency is to direct you to managers, administrators, or supervisors. Yet, the farther away we are from the "real" activity, the more abstract the descriptions of "work" become. Managers and supervisors can be very useful in helping you select performers with whom to work, and may be good sources of information for an interview. However, they cannot provide the level of detail you need for an observation session.

Even if the client understands that you need to observe performers, they still may not select the right ones. We all want to put our best foot forward. Thus, it is not surprising that clients want to schedule your sessions with expert performers. As discussed in chapter 3, we prefer a mix of performers—experts, with many years of service, and novices—because it helps to understand how performance may differ based on length of service or related factors.

Clearly communicate your requirements for selecting performers to observe to your client. The following questions may help you work with your client to select performers:

- Are these people currently performing the job, or part of the job, for which the system we are designing will be used?
- As we look across the total work processes that the system will support, do we have representative performers from each area?
- Of the selected performers, do they represent experts and novices to enable us to get a better picture of the activities performed and how they may vary according to performer and expertise?
- Would these people be able to provide us with examples of actual scenarios, forms used, system screens used, and work artifacts produced?

Planning the Sessions and Communicating With Performers

Interactive observation sessions yield the best results when you carefully plan and prepare for them. The next sections highlight activities that will enhance your success with these sessions.

Communicating the Goals and Purpose of the Session. As discussed in chapter 3, the effectiveness of an elicitation session depends in part on how the session is planned and communicated to the customer and performers who will be involved. Preparation includes:

- Determining session purpose and goals.

- Communicating the need for the session and motivating performers to want to participate.[3]
- Explanation of the user-centered approach.
- What analysts expect from performers.
- Session logistics (where, when, etc.).

Creating a Session Schedule. Performers should also receive a schedule that lets them know when the analysts will be observing them, and the place where they will meet. We are most comfortable with sessions that run approximately 1½ hours in length because of human attention spans and memory constraints. However, the goals of your session, and the type and duration of activities to be observed, should determine the length of the session.

As discussed in chapter 3, plan the schedule carefully, leaving some lag time. Review it with the customer, then produce and disseminate it to the performers you will observe.

CONDUCTING THE INTERACTIVE OBSERVATION SESSION

In chapter 7, we note the importance of "active listening" during the interview, which requires complete attention and considerable cognitive activity on the analyst's part. The same is true for interactive observation sessions. The analyst must be completely attuned to the activity being performed. You must grasp the "big picture" and the scenario being played out. You must be able to categorize the activities in real time. You must actively form hypotheses about what is happening and be able to ask questions to confirm or deny your hunches. You must listen carefully for important clues to information you might want to pursue during the follow-up question period. You also are attempting to note discrepancies between what you were led to believe would occur and what you are actually seeing. Finally, you are managing some type of notetaking endeavor to log what you see and your initial impressions, as well as to provide memory joggers that will enable you to ask useful follow-up questions.

Anatomy of Interactive Observation Sessions

The body of an interactive observation session is comprised of observing the activities being performed, handling follow-up questions, keeping the session on track, and taking notes. The next sections describe each of these in more detail.

---

[3]A sample memo to performers appears in chapter 4.

Observing Performers

The interactive observation session is focused on what the performers really do, within the context of the actual job environment. Thus, we must be able to capture "what they do" in such a way that we can make sense of it. We believe it is possible to do this only when we combine observation with questioning techniques. We do not believe a full structured interview should be conducted in combination with observation. Experience has proven that it takes observation, followed by questions that aid us in understanding what we have seen, to yield efficient information.

How and when questions occur is of paramount importance. If you interrupt a real-life activity or event to ask questions, you have impacted the normalcy of that event. (See chapter 10 on process tracing for more on this topic.) In addition, if you interrupt the event, you risk defacing a scenario that you can later decompose and analyze in further detail. Thus, most analysts agree that it is more prudent to observe the activity first to see what occurs, and make recordings of the performance, work flow, and extenuating circumstances. Query the performer after the event has concluded (or their activity in the event is complete).

During the observation session, you must also note what equipment, tools, references, or information is used, and at what point in the process it is used. Prior to each session, we construct a worksheet that we can use to work with each of our selected performers. This worksheet provides us a way to document equipment and tools associated with each observed activity. It enables us to go back and ask about these devices after the activity is complete, and to begin our review and analysis activities. Figure 6.5 is an example of an observation worksheet. The observation worksheet is an organizer. If you find it easier to take notes freeform, or in a graphical flowchart format during the session, you can always transcribe the information you capture onto the worksheet at a later time.

Finally, the analyst must look at the observed activity in terms of "What went wrong?" Identify problems that occurred during the performance of an activity. Query the performer about what happened, what they believe the cause might be, and what could be done to eliminate the problem or reduce its impact.[4]

Handling Follow-up Questions

After the activity has been completed, you should be able to consult your notes and ask follow-up questions. The intent of these questions is to:

---

[4]Interpret a performer's description of the cause of performance problems carefully. It is a human tendency to rationalize one's own performance. Additionally, people often confuse causation and correlation, reporting factors that were present at the time of the error that may not have actually contributed to or caused the error.

Performer Name _____						
Probable system interaction:						
___provide information ___review information ___create reports for others						
List tasks/activities related to the goals of the Trauma Registry System:						
Activity	Context/ location	Information provided	Information used	Resources— people, forms, things, systems	Key problems

FIG. 6.5. Observation worksheet to document activity, and resources or equipment used.

- Clarify and better understand what you saw, or what you heard the performer say during the activity.
- Discover what factors triggered the activity or response you witnessed.
- Understand protocol or procedures that were demonstrated or ignored during the activity.
- Discover specific details about the activity, such as inputs, outputs, key values, critical success factors that were stimulating actions, motivational issues that may have driven performance, training and coaching that is required to achieve that level of performance, etc.

Thus, you are listening for what they say that helps you gain a better view of their task. This includes talking to coworkers, customers, vendors, and possibly themselves. It also may include directing subordinates to complete parts of the activity. As you hear what they say, you are listening for communication intent, as well as content. The intent of a comment may have been to gain information that was necessary to the activity. On the other hand, the intent of the communication may have been polite, political, and not directly purposeful to meeting the goal of the activity. Part of your job is to comprehend the content of what they say, and the other part is to discover intent. In highly interactive environments, this may require you to keep a matrix that helps you associate communications, what appeared to be the stimulus for the communication, at whom the communication was directed, and for what possible purposes.

Finally, your job also includes listening to what performers say *after* they have completed the activity, in response to your questions. Of the many different types of questions analysts could ask, the ones most appropriate to the interactive observation include those shown in Table 6.1.

TABLE 6.1
Types of Questions for Interactive Observation

Questioning for:	Example
Stimuli and cues	"What did you hear that made you decide that the device was not an allowable commodity?"
Facts and knowledge	"What information, policies, or procedures led you to turn down the shipment?"
	"I saw you consult that manual (or other job aid) as you completed the activity. What section or pages were you viewing? Why did you consult it at the point you did?"
Goals	"What primary goal were you trying to reach?"
Constraints	"Were you pressured or constrained by factors, such as time?"
Scenarios or Cases	"Did this activity remind you of another case or event that helped you know what to do next?"
Hypothetical World	"What if you had access to this type of information? How would that affect the process?"
Verification	"It looked like you really used only 7 of the 15 available fields on this screen. Is that correct? Is this typical, or do some events require all 15 fields?"
Process	"Who uses the information you entered during this event?"

Keeping the Session on Track

It is not uncommon for the analysts or performers to become sidetracked by other aspects of the job that may not relate directly to the goals for the session. You must manage interactions and progress in interactive observation sessions. This requires a clear understanding of the goals for the session and a decision as to the depth of information required. The analyst must accomplish these management tasks during the interactive observation session:

- Recognizing how much information has been revealed.
- Identifying areas or pockets of missing information.
- Understanding the observed process or task on a number of levels, such as:
  What activities made up the observed incident?
  What goals had the performer been working toward?
  What constraints were in place to stress the process?
  What performance problems occurred?
- Recognizing when performer comments are not relevant, and interceding to bring the session back on track.
- Knowing when you have seen enough to accomplish session goals.

Taking Notes and Recording Information

How you take notes enables or restricts how you can analyze the information you have gathered (Blomberg, Giacomi, Mosher, & Swenton-Wall, 1993). If there is only one analyst in the session, it may be difficult to capture all necessary information using handwritten notes. The fast-paced job environment will render it nearly impossible to record all of the information needed. For this reason, we frequently team analysts for interactive observation sessions. One analyst is responsible for observing and querying the performer and may take transient, temporary notes to jog his or her memory. The second analyst is responsible for managing the recording of the information, using sampling worksheets, video taping, or audio taping.

If annotating by hand is the only mechanism used, analysts must decide how the information will be captured. Will you capture information exactly as it is stated, or will you summarize and paraphrase? Will you capture paralingual responses (e.g., grunts and partial language)? How will you indicate what is said in a way that can be linked to what is done?

Video recording is gaining in popularity and use in interactive observation sessions because it enables analysts to record the actual job environment, as well as the activity that the performer completed. It provides a fairly complete record of both verbal and nonverbal information. It frees the analyst to concentrate on the activity and the queries needed to understand it better. While analysts may still use notetaking to help organize themselves and remember information they want to target during questioning, the video recording can serve as a supplement to these notes. Later, it can be viewed to review important information, correct erroneous characterizations and interpretations (Suchman & Trigg, 1990), search for missing information, or gain a better understanding of the work process. However, videotaping is not without its drawbacks, most notably:

- Videotaping is obtrusive. It can be distracting, even to experts (Hart, 1986). Furthermore, some performers may respond differently (i.e., less like what they actually do) because of fear of reprisal if standard procedures aren't followed.
- Reviewing videotapes and summarizing findings is time consuming (McGraw & Harbison-Briggs, 1989). Our experience indicates that it can take two to three times as long as the session itself to do a complete review and transcription of a session tape.
- Some workplaces and states may have restrictions on videotaping, or special procedures that must be followed if taping is used. For example, one client did not allow us to record because of a law that required the organization to notify a customer that the call was being recorded.

This notification would have required the approval of other departments (e.g., legal) and would have delayed our project considerably.

Audio recording may be a viable alternative to video recording during interactive observations. The purpose of the audio tape is to record verbatim responses of the performer, primarily during the analyst's questioning period immediately after an observation. While the audio tape does not provide a visual history of the work environment or activity, it does free the analyst to concentrate more on the activity, and less on note taking. Later, as the analyst summarizes the session and documents the activities, the audio tape can provide a valuable set of session documentation. It also can lessen the introduction of analyst error that may occur when rewording or paraphrasing performer response using only handwritten notes.

Gathering Relics and Artifacts

During an interactive observation session, we have found it useful to think of ourselves as detectives, performance engineers, and anthropologists. Each of these professions requires a thorough understanding of the target of the observation. Each of them is ruthless in gathering concrete samples of relevant relics and artifacts.

In the memo that notifies them of the observation session, we tell performers we are interested in gathering sample forms, screen printouts, pages from reference materials, samples of work output or artifacts, and other relevant material. We reiterate this in our personal introduction of the session. We follow up by asking if we can get a copy of something just used by the performer. When we request a form, screen print, or artifact, we write it on a log. At the end of the session, compare what you have gathered to what is on your log. If there are missing samples, ask for them, or ask if the performer can copy and send them to you. Then follow up to ensure that you receive key samples.

To help organize ourselves, we label all samples and artifacts to indicate the session in which they were acquired. We keep these samples in our project's knowledge repository for later reference and use.

ANALYZING THE OUTPUT

Making sense of the output requires that we interpret what we have seen, analyze and synthesize it, and create intermediate representations that help us communicate what we have discovered with others on the development team.

Interpreting Session Output

Seeing is not believing. That is, just because we think we understood what we saw, we cannot assume that our interpretation and categorization of the activity was correct. We have viewed the job environment through our own eyes, using a mental model of the domain very unlike that used by seasoned performers. A very observable event may be relatively unimportant, whereas a very brief occurrence (that was difficult to observe) might be critical to the activity. Thus, the way we categorize the activities and interpret what we see is of optimum importance. This requires that we adhere to the guidelines that follow:

- Complete domain familiarization, prior to interpreting activities, to ensure that you can identify and label key activities when they occur.
- Use a worksheet or matrix to help keep track of what you are seeing, and the flow of activities in relation to each other, during your observation.
- Tell the performer you would like them to treat you as they would an apprentice—you are eager and well motivated, and know something about the domain, but recognize that you don't have the skills they do.
- Bury your ego and be willing to ask questions to clarify your understanding and interpretation of an event. (See chapter 7 for tips on reflective questioning.)
- After the session is over, meet with other analysts who have observed other performers, and talk through what you saw, what you learned, and how you interpreted the event.

Group Analysis Meetings

Group analysis is necessary only if multiple analysts conducted the interactive observation sessions. In most complex systems, this is the case. After the sessions are complete, the analysts should write up summaries of their transcripts and notate or code each form, work artifact, or screen print they gathered according to its source. Bringing these materials together, the analysts meet to share information, identify areas of misunderstanding or conflict, begin compiling or adding to scenarios, and merge and synthesize data from the observations.

We approach the task of merging and synthesizing data from the perspective of system goals or areas of functionality. Synthesis sessions work best if an agenda is set, defining the amount of time that will be spent initially on each area. For example, an analysis session agenda might target four major functions, spending 1.5 hours on each function. As each function is being discussed, all analysts share the information they gathered relative to that functional topic. One analyst facilitates the session. The goal is to begin synthesizing findings such as:

- Relevant work activity or goal within the functional area.
- For each work activity, identify the following:
  Who the performers are.
  General description of current way the work activity is accomplished.
  Key problems noted.
  Opportunities for revision, reengineering, or revised functionality.
  Additional notes or comments.

These findings may vary by project, depending on what you are observing. Note that the goal is not to find specific, detailed solutions, but to zero in on areas that may provide major opportunities for reengineering, further analysis, return on investment, etc. To assist us in compiling these findings, we organize them in a table format like that shown in Fig. 6.6.

After we have compiled our observation findings in this manner, we usually present our general findings to the client for discussion. We suggest that you conduct a focus group with a sampling of the performers whom you met and observed (see chapter 11 for details). A group session will give you an opportunity to revise any misconceptions, identify areas in which performer groups agree or disagree, and estimate the areas in which your system has the potential for major impacts (i.e., enhanced process, return on investment opportunity, etc.).

Creating and Using the Interactive Observation Session Output

One way to use the output from interactive observation sessions is to construct activity sequence tables and charts that represent the sequence of observed activities. These charts are useful when your goal is to document

Functional Area/Activity	Performed by	General Description of Current Problems	Opportunities or Suggestions for Revising Activity	Notes

FIG. 6.6. Sample summary matrix to synthesize observation findings across a group of analysts.

the sequence of activities, and how various activities are related to each other. They are especially helpful in identifying primary activity paths, or sets of activities that consistently follow each other. They also help identify optional paths. This information can be used as input to the design of the user interface, and in identifying more optimal activity paths through reengineering. They can provide input into the process that results in revised scenarios, which would demonstrate how revised activity paths enhance the work efficiency or accuracy.

Completing Activity Sequence Tables and Charts

Review the observation worksheets you compiled for a given activity area.[5] Synthesize other worksheet information that may have been recorded by other analysts on your team. Using a tabular format, plot an activity and indicate the activity that followed it, as shown in Fig. 6.7. As indicated, 14 calls were observed. The performers, Smith and Rogers, each handled 7 calls. When a call comes in, the performer must identify which activity is most appropriate to the caller's needs. The first activity performed (i.e., Dispatch in each case) is documented. If the caller requests another type of assistance, the performer completes the appropriate activity to meet that request. In our example, other activities included ordering supplies, calculating a rate, tracing an item, and setting up an account. Two of the calls were completed with just one activity (see calls 3 and 7).

Another way to analyze the sequence of activities is as a flow chart annotated to illustrate how often a particular path was followed across all observed performers (i.e., relative frequencies). Construct a flow chart so that the first activity whose paths you want to examine is shown in a box. Create additional boxes to indicate the subsequent activities that were observed for this activity. Draw lines connecting the activities. Next, either review the tabular format you created (see Fig. 6.7), or count the total number of times this activity was observed. Then count the number of times each path was taken to a subsequent activity. Representing this number as a ratio, place the tabulation on the line that links activities. This should illustrate the proportion of the time a particular activity followed another. Figure 6.8 is an example illustrating the sequence of activities and optional activity paths for the data represented in Fig. 6.7. If the first activity is not always the one shown on the left in Fig. 6.8, it may be useful to construct other graphs from the perspective of the other activities that sometimes occurred first. For example, in a call center application any of the activities shown could be the first activity (based on the needs of the caller), with the exception of "None," which indicates the call is over.

---

[5]Note, you can construct a separate table or chart for each performer to illustrate variance among them, but we prefer to synthesize observations into a master table, or chart.

Call #	Performer	First Activity	Next Activity
1	Smith	Dispatch	Supplies
2	Smith	Dispatch	Rating
3	Smith	Dispatch	None (next call)
4	Smith	Dispatch	Tracing
5	Smith	Dispatch	Supplies
6	Smith	Dispatch	Supplies
7	Smith	Dispatch	None (next call)
8	Rogers	Dispatch	Rating
9	Rogers	Dispatch	Supplies
10	Rogers	Dispatch	Tracing
11	Rogers	Dispatch	Acct. set up
12	Rogers	Dispatch	Supplies
13	Rogers	Dispatch	Rating
14	Rogers	Dispatch	Rating

FIG. 6.7. Summary of activities with subsequent activity shown.

Representing Variance Among Performers

Some observations may reveal variances among the ways performers complete activities, or even variances among different episodes of a single performer's activity. Variance is not negative, but should be captured and analyzed to determine how the performances vary, and to attempt to define why they differ. You may discover that although all of the activities have the same goals, there are many different and acceptable ways to achieve them. Or it may become clear that experts and novices perform the task differently, even though both reach the goal within acceptable time and accuracy ranges. Finally, it is always possible to identify through your observation a variance the client did not expect you to find—new and perhaps (albeit unsanctioned) better ways of completing activities.

The goal of this type of analysis is to highlight key differences and to target activities that vary significantly among performers. The activity may vary according to its goal, starting point, ordering of major processes, or output. The purpose of this activity is to compare variance at a high level, focusing on major activities, not to compare task- and subtask-level differences, which can come later, after more complete task analysis.

Identify the observed activity that revealed variance significant enough to merit its presentation and initial analysis. Enter the name representing the observed activity, along with any specific details on the scenario observed. You may find that only slight differences in characteristics of each scenario stimulated the variance you observed.

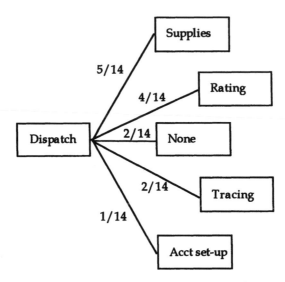

FIG. 6.8. Occurrences of each sequence are noted as a ratio on the line linking activities. The ratios enable the analysis of relative frequencies.

Graph the activity, or use textual descriptions to compare and contrast the variance. Do not yet attempt to create complete work flows at task levels—confine your presentation to the top level. Figure 6.9 illustrates one format that can be used to present variance.

ADVANTAGES AND DISADVANTAGES
OF INTERACTIVE OBSERVATION SESSIONS

Interactive observation is a popular technique that is quite effective in helping analysts ground themselves in the job context, begin to develop a mental model for the activities performed, and become acquainted with performers. It is considered more valid than the interview because simply being in the workplace with the performer demonstrating job activity yields higher task fidelity. The task of demonstrating an activity by actually doing it yields data in which we are more confident than simply having someone describe what he or she does. The focus is on *what* they do, not their description of how they follow procedures (which they may not normally follow). Watching performers complete an activity also enables analysts to verify what they did and whether it worked or met the activity's goal. Furthermore, as we observe activity, we can see the forms, screens, and other resources used, and can gather examples of these tools, as well as work artifacts produced.

Observed Activity: _____

Performer 1	Performer 2	Performer 3	Performer 4
Scenario observed:	*Scenario observed:*	*Scenario observed:*	*Scenario observed:*
Activity flow:	*Activity flow:*	*Activity flow:*	*Activity flow:*

FIG. 6.9. A format to assist in illustrating variance among performances. Content can be textual summaries or flow charts.

From the performer's perspective there are also many advantages. It is easier for them to do the activity, in most cases, than talk about it. Additionally, the workplace "anchors" the performers; what they see and touch stimulates their memory, making it easier to recall activities, vocabulary used to describe it, and ideas for improvement. Finally, observation enables us to see how the performer interacts with other performers and managers when problems arise or support is required.

Interactive observation is not without disadvantages. First, there may be domains or activities for which observation is not the most appropriate technique. For example, if the process or activity is not easily observed, session output may be performer descriptions of the task. Second, the analysts may not have enough background in the domain to recognize what they are seeing and whether it is important or not. Third, although observation may reveal the activities that occur, and enable us to document their duration and frequency, analysts can easily misinterpret this information. For example, you may assume that an activity that has a long duration is more important than short ones, which may not be valid. Furthermore, shorter activities may be harder for the analyst to recognize and discriminate.

In addition, analysts must recognize that an activity may look different when completed by marginal performers compared to the same activity performed by experts. Furthermore, what you observed may not adequately represent the total work, or may be specific to particular days of the week, shift times, seasons, or regions of the country. If this is the case you must ensure that sampling occurs across performers, numerous sites, days, or times. And regardless of who we observed, creating documentation of sessions that describe actual events may involve tape transcriptions or summarization that is time consuming to complete.

After a series of observations has been conducted, analysts should com-pare what they saw to the overall goals of requirements elicitation and domain analysis to identify areas that were not observed. Once identified, analysts must put a plan in place to use observation or other techniques to elicit information to meet outstanding goals.

Finally, analysts and performers both must realize that observation helps reveal the *current* processes in place. We will not seek to replicate them, but to understand them well enough to build on them, the work context, and the performer's mental models as we design our new system or interface.

Chapter 7

Conducting and Using the Interview Effectively

The interview is used more often than any other technique to elicit domain knowledge and requirements from an expert (Gammack & Young, 1985). In research studies that measured and compared the output of different elicitation techniques, the structured interview provided 30% to 35% coverage of a procedural domain; other techniques provided 8% to 28% coverage (Shadbolt & Burton, 1989).

Not every interview is effective, however. The ability to conduct interviews that yield desired information at the right level of specificity varies widely among analysts and engineers. More than any other technique, it demands a high level of questioning, listening, facilitation, and mental organization skills. Additionally, it is often considered the "default" technique and used when another technique might have been more effective. Consequently, it has received mixed reviews. For those who select it wisely and conduct it well, it is an extremely effective technique. Used in combination with other techniques, it becomes even more powerful.

In this chapter we define the structured interview and differentiate it from unstructured sessions as we examine its purpose and goals. Next, we provide guidelines on how to use the structured interview effectively throughout the SEP approach, including how to prepare for, conduct, and close the interview. We address verbal and nonverbal communication, feedback, questioning and listening skills, and note taking. Finally, we provide tips for managing common problem situations.

The purpose of the structured interview is to gather a wide range of information in the early phases of a project. Later, it can be used to gather specific details or to verify and validate information already gathered. It can be used at any point in the requirements gathering process or during design feedback sessions. It is a comprehensive technique that can be used alone or in combination with observation sessions, task analyses, concept analyses, or after a decision process tracing session.

The goals of the structured interview are to:

- Gather declarative knowledge (i.e., visual/verbal knowledge we can talk about). People can verbalize their job goals and the tools they use.
- Enable us to gather general information about general work processes.
- Enable us to ask follow-up, targeted questions after observation or focus group sessions.
- Gather specific information quickly, while retaining control of a session.
- Identify other information that is important to elicit, but not appropriate for the interview, to enable us to target that information for later sessions using other techniques.

UNSTRUCTURED VERSUS STRUCTURED INTERVIEWS

Unstructured interviews are often the primary vehicle for many requirements gathering activities. The analyst meets with and talks to a series of performers or interviewees. These interviews are unfocused and usually consist of the performer sharing his or her view of work, interspersed with spontaneous questions from the analyst. The interview wanders from topic to topic, focusing on an area only as it piques the interest of the analyst or elicits an emotion from the performer. If the analyst is lucky, and already knows a lot about the domain, he or she may come away with some information that will be useful. If the analyst is not well-schooled in the domain, however, he or she may miss important questioning areas or may fail to follow up on a topic to gain more depth. Because the session seems unfocused, the performer may take control of the session, leaving the analyst in a passive, "anything goes" mode of control. Consequently, unstructured interviewing seldom produces complete or well-organized output. Data acquired from the unstructured interview often seems unrelated, and exists at varying levels of complexity. This makes it difficult for the analyst to review, interpret, and integrate the content of the interview. Finally, performers rarely view unstructured interviews as time well spent.

Structured interviewing is an organized, carefully managed communication between analyst and performer. The structured interview has been described as goal-oriented (McGraw & Harbison-Briggs, 1989; Waldron,

1986). The goals used to plan the interview and, later, to keep it on track, reduce the interpretation problems inherent in unstructured interviews. Goals also help the analyst/interviewer reduce the distortion caused by his or her subjectivity (Waldron, 1986). Additionally, the kinds of questions asked in a structured interview, and the way they are asked, force the expert to be systematic (Hoffman, 1987).

How to Use the Structured Interview Effectively

Structured interviewing is used throughout SEP to clarify, extend, and gain a more complete understanding of a topic. The tasks that follow summarize the basic guidelines we suggest. Subsequent sections in this chapter examine many of these in more detail.

1. The analyst reviews scenarios that have been generated, session write-ups produced from other techniques, or current definitions of components and devices required to support the domain. The goal is to identify areas in which specific information of a declarative nature needs to be further explicated.
2. After specific areas have been targeted, the analyst searches for gaps in his or her understanding of a system, process, or entity that would render the definition of requirements or specifications incomplete. Gaps become major question areas or topics for structured interviews.
3. The analyst works with team members to refine the plan for a structured interview session. This includes agreeing on the purpose and desired outcome for the session, compiling a list of sample questions or question areas, and suggesting which performer would be able to provide the information.
4. The analyst works with the client organization to schedule the interview, identify an appropriate location for the interview, and negotiate for the preferred performer or interviewee.
5. The analyst prepares a planning form for the session and sends it to the interviewee prior to the session.
6. During the interview the analyst reviews the agenda, then manages the introduction, body, and closing segments.
7. After the interview the analyst reviews session information, prepares a session summary, and creates any intermediate products the session generates. The analyst sends the session summary and/or products to the interviewee for review and refinement.
8. The analyst identifies additional topics in the same area that may require further investigation.

PREPARING FOR THE STRUCTURED INTERVIEW

Many people confuse the structured interview with a session in which the analyst meets and talks to a performer or domain expert, using a set of loosely defined goals. Actually, an effective structured interview requires careful planning and preparation to ensure that the analyst can meet the goals of the session. Pre-interview activities include identifying the degree of structure desired and specific objectives for the interview. Next, the analyst should define the specific targets of the interview. Additionally, most analysts find it useful to predefine sample questions to help them stay on track, and to help communicate the desired focus to the interviewee.

Setting Objectives

The analyst should determine the degree of structure required and the objectives the interview should meet. For example, will the interview be highly structured, with questions posed to verify or validate information and heuristics already gathered? If the objectives are "to validate the tasks required to support the diagnosis function," you may ask primarily closed questions, seeking *yes* or *no* answers. Clarification and expansion of an issue will be required only when the answer is not what was expected. This type of session requires that you identify specific questions beforehand, and an optimum sequence for presenting them.

If the main objective of your session is to seek information rather than validate it, the session could be moderately structured. If the session objective is to seek validation *and* verification, it should be even more structured. For example, if the main object is to "elicit details about the definition and use of products" in the domain, the questions you ask will be a mix of open and closed questions. Question sets (i.e., groups of questions whose purpose is to elicit information about a specific target) may begin with an open question and end with a more closed question. Even for moderately structured interviews you should be able to identify specific targets and sample questions for each target area.

Identifying Interview Targets

After you have identified interview objectives and determined how structured the interview should be, you should define the primary interview subtopic(s) you wish to address. Many analysts make the mistake of overplanning—that is, defining more subtopics than can possibly be covered adequately in a single interview session lasting approximately an hour.[1] If

---

[1]Our experience has shown that an interview time of 1 hour to 1 hour and 15 minutes is optimum. Longer sessions are often counterproductive, due to limits on human attention span and interest.

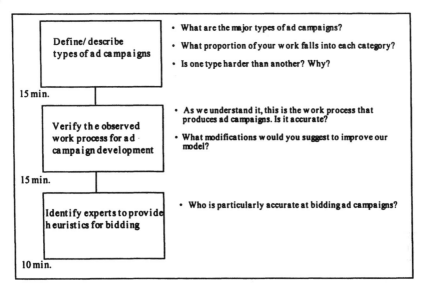

FIG. 7.1.   Interview planning map to help identify subtopics and question sets.

analysts define too many subtopics for a single interview session, it is likely that they will not all be met. Furthermore, the interviewee may perceive the analyst/interviewer as naive about the domain. To help plan sessions more realistically, we advise new analysts to actually write an outline or graph of the interview. Assuming that the interview is an hour, for example, we might decompose our interview goal into four major question sets or subtopics. We would think of each question set as consuming approximately 10 minutes (planning for opening and closing time). Next, we would write out two or three primary questions for each area, knowing that each of these primary questions could generate two or more secondary (i.e., follow-up) questions. If we have more questions than can fit our interview map, we need to plan more sessions, rather than attempting to cover too much territory in a single session. Figure 7.1 is an example of one of our interview planning maps for the body of a structured interview.

After the analyst has defined and limited the coverage for a structured interview, it becomes relatively easy to select the most appropriate expert or interviewee for the session. One practical tip is to ensure that the customer or client is providing the right person to fill the role of interviewee.[2] If you can articulate the objectives, subtopics, and sample question areas well, you will increase the likelihood of being assigned someone who knows the area. When the client suggests someone for the session, we ask a few questions to help confirm his or her appropriateness:

---

[2]We suggest requesting interviewees with intermediate and expert levels of skill.

- What is this person's current responsibility (managerial, individual performer, team leader) in the area being investigated?

  Sometimes clients will provide a person who is in management (for political reasons), when you may actually need an individual performer.

- Are you confident this person will be able to accurately articulate "X" at the level of detail you need for this project?

  You are reminding your client that the project's success depends on the accuracy and clarity of requirements or other information you are expecting this person to be able to elucidate.

- Will this person be available for follow-up sessions, should they be necessary?

  If you find the interviewee particularly knowledgeable, analytical, or communicative, you may want to meet with him or her again, either for follow-up interview sessions, or for other sessions.

Creating an Interview Session Plan

To create an interview session plan, transfer the interview planning map content onto a session plan form. We create such a plan form for each session we do, but it is particularly useful for a structured interview. Our experience has shown that by providing explicitly stated goals and subtopics along with sample questions, we can enhance the quality of the session.

Not only does the session plan form help prepare the analyst to conduct the session, it also helps prepare the interviewee. Most of them are concerned about the goals, subtopics, and depth of coverage they can anticipate during the interview. After you have created the session plan, fax or e-mail it to the interviewee for review. He or she can reflect on the subtopic areas, consider the sample questions, and actually begin the process of mental data gathering before the interview begins. This pre-interview mental preparation is critical, because the interview yields information that is of a declarative nature. The expert must be able to retrieve it from memory structures and be able to talk about it. If the interviewee has the session plan a few days before it is conducted, he or she is more likely to provide the type of information you require with less frustration.

CONDUCTING AN EFFECTIVE INTERVIEW

An effective interview is comprised of three major parts, each of which is critical to the overall success of the interview: the introduction, the body, and the closing. The following sections detail each of these, describe the type of activities, and provide guidelines for their completion.

Introducing the Interview

The introduction or opening of an interview is brief, but critical. According to Zunin, the major function of this phase is to motivate the participants to communicate actively (Zunin & Zunin, 1975). What the analyst says and does during this time sets the tone for all that follows, and requires training and experience in communication and interviewing. You should appear confident and in control as the introduction begins. You must quickly build some degree of rapport with the interviewee. The tone you convey should be perceived as professional, nonthreatening, and relaxed. To convey otherwise could discourage the interviewee from participating freely and openly, or from taking part in subsequent sessions.

If you do *not* seem professional, confident, and skilled in communicating and building rapport, the project may not be able to gain access to other critical interviewees, particularly those at high levels of an organization.

Stewart and Cash (1985) suggested a two-step introduction process that is appropriate for requirements gathering. The order in which these steps are completed may vary, depending on personal preference, the relationship between the interview participants, and the specific situation.

Building Rapport

The first step is establishing rapport and trust between the analyst and interviewee. Establishing rapport involves both verbal and nonverbal messages. Verbal messages are usually of the following types:

- Introductions, using names as opposed to titles:
  "Hello, I'm Karen Smithson, from Cognitive Technologies. You're Dr. Minion?"
- Attempts to deal with any initial anxiety on the interviewee's part:
  "I've been looking forward to talking to you about this project, to help us to build a system that will support your needs. Did you receive the interview plan for the session?"
- Personal inquiries or "small talk":
  "I'm told you are writing a text on the diagnosis of depression." (We try to minimize this type of message to ensure it does not distract from the purpose of the interview.)

Session Orientation

The second step in the introduction is to orient the interviewee to the session's purpose and goals. This involves providing an opening statement describing why you requested the session and its purpose. If the interviewee is not fully aware of the project that the interview supports, provide a summary-level description of it.[3] The purpose of this step is to ensure the interviewee that the session will be relevant and that his or her participation has a bearing on the project's success.

After you complete the general overview, present the agenda for the interview session. If the session is held in a conference room, we suggest posting the interview agenda on a white board or flip chart. Otherwise, show the interview plan form to the interviewee. Point out the main question areas, using appropriate domain vocabulary, and briefly describe the approximate amount of time you expect to spend in each area.

Warning—during this part of the interview it is critical that you appear organized and in charge. We have seen interview sessions degenerate at this point because the interviewee perceives that the analyst is not confident, knowledgeable about the domain (and its vocabulary), or capable of leading the activity.

Conducting the Body of the Interview

The body of the interview begins as soon as the introduction is complete and should reflect the agenda and goals on the session planning form. In fact, we keep this form on the table in front of us during the interview to motivate us to stay on track. If the interview takes an unwanted detour, we can point to the agenda to help us regain control.

All interviews have some type of structure and require effective verbal and nonverbal communication. During the body of the interview the analyst will ask questions in a way that produces value-added communication. Additionally, the analyst will apply active listening techniques that assist in determining follow-up questions. The analyst must also monitor nonverbal communication and be alert to instances in which the verbal and nonverbal messages don't seem to be consistent. The analyst will use leadership and interview management techniques to keep the interview body on track, progressing toward the goals. Finally, the analyst must ensure that the interviewee continues to feel comfortable and effective. The following sections describe these activities in more detail.

Interview Structure

Various communications specialists describe techniques that interviewers and analysts use to structure and connect the questions that comprise the body of an interview. Any of these connective techniques may describe an entire structured interview, or may be used within specific subsets (e.g., to support a question set) of a structured interview.

---

[3]We call this our "marketing statement." It is a quick, no-nonsense description of the project and how it might benefit the interviewee.

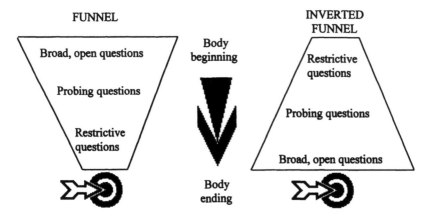

FIG. 7.2. Two interview body structures.

Sequencing techniques that we have found most appropriate include the funnel and inverted funnel technique, as depicted in Fig. 7.2. In most interviews we conduct, we tend to use more than one of the selected structures—one for each question set or subtopic on our agenda. Thus, the interview becomes a series of connected funnels, in which each funnel represents a question set. The skilled interviewer may even combine both of these structures during a single interview.

Funnel Structure. The funnel is the most commonly used structure for structured interview sessions. It begins with broad, open questions, continues with probing or focusing questions, and concludes with more restrictive, closed questions. The funnel structure can be very effective for several reasons. First, because it begins with open questions, it is easier for the analyst to establish rapport, because open questions are less threatening. Second, beginning with open questions enables the analyst to evaluate responses and refine secondary follow-up, or probing, questions. This maneuver allows the analyst to discard intended questions that may no longer be necessary because the interviewee has already volunteered information in response to an open question. Finally, through the responses the interviewee makes to the initial, broad questions, the analyst may discover that the interviewee is *not* a good match for the session topics.

Inverted Funnel. When using the inverted funnel, the analyst begins with restrictive, closed questions and offers less restrictive ones as the interview progresses. This sequence enables the analyst to begin with a specific, well-focused question, which can have a positive impact on the interviewee. First, he or she sees that the session is well-defined and is more likely to perceive the session as business-like. Second, this type of question enables the interviewee to refresh his or her memory, setting the stage for the topics to be discussed. Third, this structure allows the analyst to use closed questions to draw out an interviewee and end with a generalization or summary statement (Stewart & Cash, 1985). Finally, this type of structure is very useful if the analyst has reason to expect that the interviewee will be hard to manage.

Using Verbal and Nonverbal Communication

The communication model shown in Fig. 7.3 illustrates the complexity of the communication process on which the success of the structured interview depends. The sender recognizes the idea he or she wants to communicate. Based on background, experience, concepts, vocabulary, and goals, the sender recodes the idea into a format that can be communicated to the receiver. The communication is sent, with the sender's intended meaning embedded in the words and gestures. The receiver takes in the message and decodes it, using his or her background, experience, etc. The decoded message is analyzed and hopefully, the idea is understood. The extent to which the analyst and interviewee have similar backgrounds, experiences, vocabulary, concepts, and goals or motives determines the ease with which they communicate.

Verbal communication relies on the ability to express meanings, concepts, beliefs, and rules through spoken language. This requires that the interviewee have work experience in the domain prior to the interview, and has encoded, or internalized, important concepts, vocabulary, and rules. Next, it requires that the interviewee be able to listen actively to the question posed and interpret its intent based on the goals of the session and the context of the question set. It may also require that the interviewee translate terms the analyst uses into equivalent terms with which the in-

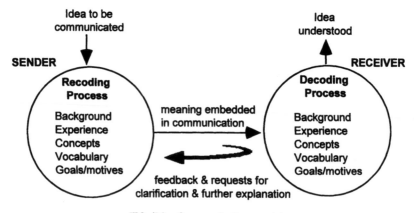

FIG. 7.3. Communication model.

terviewee is familiar. It assumes that the interviewee has organized or structured the domain concepts, meanings, and rules in ways that enable him or her to recall or retrieve them when necessary. Finally, it requires that the interviewee be able to recode these meanings, concepts, and rules into vocabulary that can be expressed in a meaningful way in response to questions. Communication problems can occur at any of these points.

On the other side of the table sits the interviewer or analyst. Effective interviewing requires numerous mental activities and skills on your part as well. First, you must be prepared for the session and should have reviewed the questions beforehand to ensure appropriate use of domain terminology. The extent to which you frame questions in the interviewee's language will improve the ease with which the interviewee responds, and the quality and effectiveness of the answer. Additionally, you must carefully construct how you frame and ask each question. Guidelines include:

1. Ask the question once, clearly. Don't try to explain what you really want, or to reword the question, unless you are positive the interviewee has not understood you.
2. Link follow-up questions to the previous response or to the main topic in the question set, when possible, to help provide a continual context for the question.
3. Beware of the complex question. Refrain from embedding subquestions within the main question. Ask one question at a time to help the interviewee formulate the most appropriate answer with the least possible confusion.
4. Refrain from bias in the way you frame or ask the question.
5. Use both verbal and nonverbal communication to provide information and to decode the messages the interviewer is sending you in response to a question.
6. Ask the right questions, at the right level. Vary your questioning style and use active listening techniques (see *Using Active Listening*) to enhance the quality of the response.

Verbal Communication

Verbal communication is the heart of the interview. However, verbal communication is imperfect. Both the interviewer/analyst and the interviewee use words to express internal meanings. The meaning one person ascribes to the words used may not be the meaning that is understood by the other party. These differences result, in part, from the varied backgrounds, experience, knowledge, and motives held by each interview participant. To decrease the impact of a disparity in the vocabulary used by the interviewee and interviewer, the interviewer/analyst should do the following:

- Study the domain vocabulary, acronyms, and concepts prior to conducting the interview.
- If the interviewee uses a word differently than expected, watch for other cues before interrupting him or her for clarification. If used judiciously, active listening techniques help to refine and clarify intended meanings.
- If the meaning of a word is still not understood or is confusing, the interviewer/analyst can ask how it is similar to, or different from, the term (and use) that was expected.

Additionally, the interviewer/analyst must address the impact of verbal ambiguity on the interpretation of responses to questions. Consider the way most of us verbally quantify information using words such as most, much, a majority, little, average, none. Each of us has his or her own meaning for these words, but few would agree on percentages that would correlate with them. If the interviewee uses the word *average* to describe a speed or the chance of something occurring, he or she has introduced verbal ambiguity to the interview. Unless the analyst is aware of this and attempts to reduce ambiguity by asking focused follow-up questions, the value of the information is questionable.

Another difficulty related to verbal communication is the fact that the two interview participants may understand a verbal message in different ways. Our ability to decode a verbal message, process it, and understand its meaning (i.e., the "deep" or semantic idea the words represent) depends on our backgrounds, vocabularies, and experiences. Even if these are similar for the interview participants, misunderstanding may still occur because other signals (such as intonation, voice, and pitch) accompany and may distort or clarify the verbal message. For example, examine Fig. 7.4. The way a speaker stresses different words in the same sentence impacts the meaning the message conveys.

Nonverbal Communication

Some of the difficulties that arise during verbal communication can be clarified if the analyst is aware of, and works to interpret, the supporting nonverbal communication. Nonverbal communication conveys meanings that may enhance, substitute, or even contradict the accompanying verbal communication. We use many tools to communicate nonverbally. Figure 7.5 depicts some of these.

In our role as interviewers, we are responsible for facilitating the session and ensuring that active, meaningful communication is occurring. This responsibility extends to nonverbal communication as well. We must be cognizant of the nonverbal communication that we use and should attempt

Message	Implication and Follow-Up
The *training* manual does not say to do that. 	**Implies:** there is some other manual that says to do that. **Follow-up:** analyst should ask the interviewee to identify the proper manual.
The training *manual* does not say to do that. 	**Implies:** some other training product or person gives those directions. **Follow-up:** analyst should ask the interviewee to identify the training product that provides those directions.
The training manual does not say to do *that*. 	**Implies:** the directions or topic being discussed is not correct. **Follow-up:** analyst should ask what the correct action is.

FIG. 7.4. Stress, intonation, voice, and pitch modify the meaning of verbal communications.

to send consistent verbal and nonverbal messages to the interviewee. Through nonverbal communication we can convey the following:

- I'm confident and in charge of the session.
- I'm interested in, or bored with, your comments.
- I see the value of that statement and want more details.
- I'm confused or have misunderstood what you are saying.
- I think I've lost control of the session.
- I'd like to move on to the next topic.
- I want to interrupt you and redirect the communication.
- The session is over.

We must be able to recognize and interpret important nonverbal communication messages that the interviewee seems to be sending. This does not imply that every body posture change, eye movement, facial expression, or paralingual (e.g., partial language) cue is meaningful. Each type of communication must be interpreted using other clues from the current

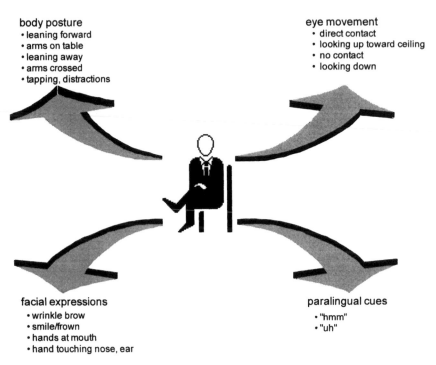

body posture
• leaning forward
• arms on table
• leaning away
• arms crossed
• tapping, distractions

eye movement
• direct contact
• looking up toward ceiling
• no contact
• looking down

facial expressions
• wrinkle brow
• smile/frown
• hands at mouth
• hand touching nose, ear

paralingual cues
• "hmm"
• "uh"

FIG. 7.5. Nonverbal communication tools.

context of the communication. Things to look for include nonverbal signs of frustration, lack of interest or motivation, confusion, and so on. If the analyst already suspects confusion, uncertainty, or mixed messages, he or she should examine relevant nonverbal communication when deciding how to respond.

Communication texts generally agree that an interviewee transmits certain nonverbal messages during an interview (Stewart & Cash, 1985; Nierenberg & Calero, 1971). For example, Table 7.1 presents commonly accepted interpretations for some examples of nonverbal communication.

Using Feedback During the Interview

Feedback is the "give and take" by which the interviewer receives, understands, and communicates that understanding has occurred or that clarification is needed. Feedback is continuous and immediate, and is expressed through both verbal and nonverbal communication.

Feedback takes one of three forms: verification, disagreement, or revision. Table 7.2 illustrates examples of generalized feedback.

Guidelines to the effective use of feedback in the interview session include the following (McGraw & Harbison-Briggs, 1989):

TABLE 7.1
Interpreting Nonverbal Communication

Nonverbal Mechanism	Common Interpretation
Leaning forward toward the other participant	Interest, eagerness
Pushing away from the other participant, table, etc.	Disinterest
	Agitation
	Time to leave
Hands around or over mouth during response	Message may not be totally accurate
Eyes moving up toward ceiling	Thinking, recalling
Removing glasses and placing ear piece in mouth	Gaining time to evaluate, seeking information
Chin stroking	Considering, decision making
Tilted head	Interest
Touching or rubbing nose	Not sure, have some doubts
"Tsk" sound	Disgust, disagreement
Clenched hands, wringing hands	On the "hot seat"
	In sensitive area
Hands together in front of body in "steeple" shape	Very confident
Tugging at ear	Wants to interrupt
Crossed arms and legs, head down	Preoccupied

TABLE 7.2
Generalized Feedback Examples

Form of Feedback	Example
Feedback that VERIFIES	"Yes, that is what I said."
	Nodding head "yes"
Feedback that states DISAGREEMENT	"No, that is not what I meant."
	Shaking head "no"
Feedback that REVISES	"Actually, what I meant is . . ."

1. Provide the interviewee with appropriate feedback to illustrate your degree of understanding.
2. Use both nonverbal and verbal prompts for clarification when needed.
3. Don't hesitate to use supportive and appreciative noises such as "uh-huh," "hmm . . ." and so on.
4. Do not attempt to provide feedback for every comment the interviewee makes; use questioning techniques (detailed in the next section) or make clarifying statements when appropriate.
5. Videotape the session when possible, both to help analyze the session and to review your performance. How are you using feedback? Does it enhance communication (e.g., making the interviewee feel positive about himself and the value of his contribution, gaining more infor-

mation and clarification)? Does it detract from the session (e.g., demonstrates your knowledge, defends your ideas, argues)?

Understanding and Using Categories of Questions

Many variables help determine the success of an interview session. One of the most important skills is the interviewer's ability to present various types of questions in a sequence that enhances the accuracy and specificity of the responses. It is imperative that the interviewer be able to use questioning techniques effectively, including selecting the most appropriate technique, formulating questions, analyzing output, using active listening, and using probes. The next sections explore types and levels of questions, and offer suggestions for improving the phrasing of interview questions.

Selecting Question Types—Open Versus Closed. Open questions are broad and do not limit the response the interviewee can make. Closed questions set limits on the type and extent of response. Both types have advantages. Most interview sessions require the use of a variety of open and closed question types with varying degrees of restrictions. Consider the following when making the determination to use open and closed questions during the session:

1. The amount of information offered by an interviewee in response to a question decreases with increased question restrictiveness.
2. The interviewer's degree of control over the session increases as the amount of information offered by the interviewee decreases.
3. The degree to which the interviewee believes he or she can add information that has not been requested increases in sessions with a good mix of open and closed questions.
4. The less the interviewer knows about the domain, the more likely that he or she will overuse closed questions in an attempt to remain "in control."

Good interviewers practice using both types of questions. Prior to conducting your first interview, run through a mock session and videotape your performance to review your questioning expertise. This will allow you to observe your ability to make a smooth transition between open and closed sessions. The following provides detail on open and closed questions to help you prepare for and use each type effectively.

Open questions place few restrictions or constraints on the responses the interviewee makes. Open questions encourage free response and are appropriate when you wish to observe high level responses to discern any of the following:

- The interviewee's scope of understanding.
- The certainty with which the interviewee responds.
- What the interviewee generally thinks is important about the topic.

Furthermore, open questions enable the interviewee to offer information that you did not know to ask for—a situation that is highly probable when you are gathering requirements in an unfamiliar domain. Trigger words for open questions include discuss, interpret, explain, evaluate, compare, if, and what if. Table 7.3 presents samples of open questions.

Answers to open questions enable you to observe and begin to understand the interviewee's use of key vocabulary and frames of reference, his or her use of concepts, and clues to the way the interviewee has organized the domain.

There are disadvantages to the indiscriminate use of open questions, however. Responses to open questions take a lot of time, but may reveal little information *at the level required* for you to identify system requirements or build domain models. If you generously use open questions you will have to work harder to control the interview. You will continually be required to focus and redirect the interviewee. You will have to follow up with probes to garner specific information at the level of detail you require. You will have to watch body language and hear what has not been said to determine whether the interviewee has omitted information that he or she thinks might be irrelevant. Finally, note taking and later activities involving coding and organizing the information will be much more cumbersome because of the lack of structure in the questions.

Closed questions set limits on the type, level, and amount of information an interviewee provides. They may include a choice of alternatives or hints to the level of reply desired. Trigger words for closed questions include who, what, when, where, and name. Table 7.4 presents samples of closed questions.

Closed questions may vary in the degree to which they are closed. A moderately closed question allows the interviewee to respond with a specific

TABLE 7.3
Samples of Open Questions

"How does that system work?"	"What do you need to know before you can decide?"
"Why did you choose this one rather than that one?"	"What do you know about the fuel system?"
"How could your training have been improved?"	"What is your general reaction to this statement?"
"Describe your work process to me."	"How do you determine who needs assistance first, second, and so on?"

TABLE 7.4
Samples of Closed Questions

"Which of these planes have you maintained?"	"How many hours in the cockpit have you had?"
"Which of these types of investments do you consider strongest?"	"Will this cause the oil pressure to be higher or lower?"
"What is the target speed or window for this maneuver?"	"Is 225 degrees the upper limit?"
"What is the first thing you do when the victim arrives?"	"What is the most difficult task to perform?"

piece of information, but does not limit the set from which he or she responds. For example, a question like the following specifies particular information, but does not limit the interviewee to select a bipolar response (e.g., yes or no):

"Which symptom led you to believe the problem was in the electrical system?"

If you are eliciting requirements, consider using moderately closed questions as follow-up probes or clarifying questions. You can use bipolar questions, which limit the respondent to one of two choices, to establish the accuracy of information already elicited.

Closed questions are most effective when the interviewer needs a specific piece of information. They help you control the progress of an interview session more effectively, make note taking and answer coding easier, and complete the interview in less time.

Disadvantages to the use of closed questions also exist. First, new interviewers are inclined to overuse this easier type of questioning, which may result in getting too little information, necessitating more follow-up questions or sessions. Second, closed questions may preclude the interviewee from volunteering important information that was not directly specified. Third, when you use more closed questions, you end up doing most of the talking. Consequently, you must have an excellent command of the vocabulary and concepts in the domain, as well as an awareness of the interviewee's abilities and frames of reference.

Using Primary and Secondary Questions. Communications researchers (Kahn & Cannell, 1982) recognize two levels of questions: primary and secondary. Primary questions are top level questions that you use to introduce topic areas or for transition to another question set. It is possible to analyze a transcription of an interview and easily identify the primary questions. They stand alone and can be understood out of the context of the rest of the interview. For example, the question, "Given these symptoms, what cause would you first be inclined to investigate?" can be understood out of context. It introduces a question set about the relative frequency of symptoms.

Secondary questions are also known as probing questions. Their purpose is to find out more about information offered in response to a primary or a preceding secondary question. They allow you to follow up on a topic area, requesting ever-more-specific information. Table 7.5 presents different types of secondary or probing questions (McGraw & Harbison-Briggs, 1989).

Interviewers must recognize the importance of secondary questions. Reviews of hours of interviews reveal that new analysts or interviewers often leave "money on the table" after the interviewee responds to a primary question. That is, they fail to follow up with appropriate probing questions that would enable them to elicit the detail needed to define requirements, components, and devices. A skilled interviewer listens to responses to previous questions, compares them to his or her expectations, analyzes meanings that have been communicated, determines if the answer is satisfactory, and generates probing questions for follow-up.

TABLE 7.5
Secondary Probes

Type of Probe	Tips for Using
Silence Probe	If the interviewee does not appear to have completed a response, remain quiet and use active listening techniques to encourage further response.
Prompting Probe	If nonverbal body language does not stimulate further response, use verbal techniques such as, "Please continue . . ."
Depth Probe	If the interviewee provides surface-level information, follow up with, "Explain further why you diagnosed it like that . . ." or "Tell me more about . . ."
Specifying/Labeling Probe	If the interviewee has provided information that is not specific enough, or at the appropriate level of detail, direct or focus the interviewee with probes such as "What does <BVR> mean?" "What value would you define as 'fast enough'?"
Reflective Probe	If you believe the interviewee has provided inaccurate information, restate the answer that was provided, emphasizing the part of the probe that contains the questionable information, such as "Do you mean MPG or KPG?" "Did you mean may not?"
Last-Chance Probe	Use when you want to make sure all relevant information has been elicited in a question set or topic area, such as "Is there anything more we should know about tracking and documenting medical complications?"

Phrasing Questions Effectively. Preplanning a good mix of open and closed questions and knowing how to use primary and secondary questions effectively will go a long way toward the elicitation of target information. However, the communication process can be impacted negatively if the selected questions are not phrased properly. Question phrasing involves three different but related elements: terminology, level, and complexity.

Use of appropriate terminology helps build a bridge between the interviewer and interviewee. Interviewers must be able to use important domain terminology in phrasing questions. If you use the language of the domain, the interviewee spends less time translating what you are asking and more time formulating thoughtful answers. Your mastery of critical domain vocabulary enhances rapport and maximizes the time allotted for the session. The following guidelines apply to the use of terminology in question phrasing (McGraw & Harbison-Briggs, 1989):

1. Avoid using words that may have multiple meanings.
2. Avoid using vague words (e.g., most, much, average).
3. Control the use of words that may express unintended synonymy (e.g., could, should).
4. Offer contexts to guard against misunderstandings caused by words that may sound similar but have very different meanings.
5. Eliminate bias in question phrasing.

The level at which a question is asked, and a response expected, also impacts question phrasing. Carefully consider the level of information you need to elicit and ensure that you are asking questions that solicit a response at that level. If you phrase questions in such a way that they require responses that are above the interviewee's level of expertise, or are outside of his or her area, it may cause embarrassment, reluctance to participate further, or lack of honesty. If you phrase questions so that they sound too elementary or seem to be beneath the interviewee's level, you may appear naive to the interviewee, who may feel that he or she has wasted valuable time.

Finally, question phrasing is impacted by complexity. In fact, most new interviewers perform poorly in terms of making sure their questions are phrased in a clear, straightforward manner. In attempts to provide context for the question, and to "help" the interviewee answer, they create obtuse, complex questions. Common phrasing errors that result in complexity include:

1. Providing too much context for the question.
2. Asking and reasking the question in multiple ways in an effort to assist the interviewee, before giving him or her a chance to respond.

3. Providing details about the type of information wanted in response.
4. Using the question as a "soapbox" to defend a view the interviewer holds.

The example shown in Fig. 7.6 illustrates the tendency toward complexity. The interviewer makes the first three of the four phrasing errors.

Using Active Listening. One of the biggest challenges facing an analyst/interviewer is to listen actively while juggling interview management, note taking, and questioning activities. Yet active listening techniques can mean the difference between an interview that yields surface level knowledge and one that produces semantic level knowledge and detail. Listening sounds easy, but 85% of people who respond to listening surveys generally rate themselves as average listeners—or worse. While we listen, we are also involved in other cognitive activities such as formulating the response we want to make next, or recalling an episode or story we would like to share. Active listening techniques help us focus on the interviewee and enable us to communicate what we believe we heard in a way that stimulates clarification and refinement.

Paraphrasing is one useful technique that not only improves listening but also gives the interviewee an opportunity to respond. To paraphrase, you must first attend to what the interviewee is saying. Then instead of moving directly into the next question or question set, restate the meaning

"I need your opinion on the parts of this display that you find most useful. Take a few minutes to look at the display and envision yourself in the situation we discussed earlier. Which part of the display of the panel is most important or critical, or would be required in emergency situations? This is so I can can analyze information flow and later, make color selections. (PAUSE) Can you rate each lettered part of the display to indicate usefulness, using a scale of 1 to 5, where 5 is most important and 1 is least important?"

FIG. 7.6. Complexity is a common phrasing error.

perceived, using your own words and phrasing. Paraphrasing is not the same as "parroting" back what the interviewee has said. It requires active mental processing on your part, as you restate the ideas. These are examples of paraphrasing:

- "As I understand it, <restate what they said in your own words>."
- "Please point out any misunderstanding on my part, but what I think you said was that <restate what they said in your own words>."
- OK. I believe you said that <restate what they said in your own words>."

In response to the paraphrase, the interviewee will either acknowledge that you have interpreted correctly, or will commence to clarify and/or extend the idea. The result—you have demonstrated that you were listening. Additionally, you now have either confirmation of accuracy, or a better understanding of the ideas being expressed.

Reflective questioning is another technique that demonstrates active listening and involves you in responding directly to the interviewee's statements. Reflective questioning allows us to "reflect back" what we think we heard or understood to elicit clarification from the interviewee. It is useful when we want to clarify a word with more than one meaning or an answer we did not expect or understand. It is also useful in eliciting further or deeper explanation than was first offered. These are examples of reflective questioning:

- "I don't get it."
- "I'm not following the logic. Can you walk me through it again?"
- "Can you explain that in more depth, or give me an example?"
- "Do you mean *primary* assessment or *secondary* assessment?"

The interviewee hears these statements and knows we are attempting to listen and understand the domain knowledge. Furthermore, each of these statements reflects our own misunderstanding, rather than implying that the interviewee described something poorly. Finally, the result is that we have a better, deeper, more accurate understanding of the domain after the interviewee responds.

Using Wait Time. Wait time is a normal convention of two-party conversations. It refers to that period of time after the interviewer has asked a question (McGraw & Harbison-Briggs, 1989). Wait time can be thought of as the interviewee's "think time." Research in other fields indicates that wait time of as little as 3 to 5 seconds after a question or response increased the length of responses and decreased the failure of an individual to respond. Interviewers are advised to remain quiet after posing a question and to guard against using nonverbal cues that might communicate impatience. Allowing the interviewee to retrieve the required information will result in a better quality answer. Remember, if the information sought is not in short term memory, the interviewee must attempt to search for, de-compile, and retrieve it from long term memory.

If the interviewee still has not responded or indicated that he or she is thinking after 7 or more seconds, you may restate the question. This takes some immediate pressure off the interviewee, while indicating that the information still is desired. The interviewee uses this additional time to retrieve and process requested information and to formulate a response.

Note Taking and Recording the Interview

Session information can be preserved in at least three ways: handwritten notes, audiotape record, or videotape record. In many cases we combine the use of handwritten notes with either the audiotape or videotape. Even when we have an audiotape or videotape in the room, we use the handwritten notes to help:

- keep track of key information the interviewee has said, for use in follow-up questions and session summaries;
- sketch process-type information or other verbalizations that lend themselves to graphics we can use in follow-up questions; and
- pay attention to the topic and refrain from losing focus during particularly long or complex sessions.

If the only recording mechanism available is handwritten notes, we suggest that a fellow analyst accompany you and act as "scribe." This frees you to concentrate on formulating the questions, keeping the session on track, keeping high-level notes, and working with the interviewee.

Using Handwritten Notes. Handwritten notes enable you to be self-sufficient during the interview; you don't have to struggle with taping equipment prior to the session or monitor it during the session. However, few of us are taught how to take efficient notes—knowing what to write down and when to do so. In addition, note taking during an interview may impose several problems on the communication process that can interfere with the interviewee's train of thought and your own concentration.

On the positive side, handwritten notes do offer control (McGraw & Harbison-Briggs, 1989). They allow you to capture specific information, rather than everything that is said. This reduces later analysis and synthesis time, as it becomes fairly simple to transcribe the session. It is also cheaper, requiring no special equipment, such as recorders and tapes. Finally, it provides great flexibility. You can use it anywhere, with no restrictions.

On the negative side, manual note taking requires that you take responsibility for both the management of the interview and the note taking activity (McGraw & Harbison-Briggs, 1989). These two are not always congruent. As you increase your focus on the note taking activity, you decrease the effectiveness with which you can concentrate on the interviewee, the sequence of questions, and the selection of good probes or follow-up questions. Note taking also interferes with nonverbal communication. If you are concentrating on taking notes (which necessitates that your head is down), you won't see or be able to respond to the interviewee's nonverbal communication. Furthermore, you are limited in the type of nonverbal feedback you can use to help the interviewee know that you understood, or are confused about a response. Additionally, if you are new to the domain, you have a tendency to write down everything that is said. This causes long pauses in the interview and reduces the effectiveness of secondary questions and probes. Finally, you must be concerned about the impact of manual note taking on the interviewee. Certain interviewees may become so preoccupied with what you are writing down that they begin interrupting, identifying information that should not be written down, and so on. The following guidelines may help you use manual note taking more effectively (McGraw & Harbison-Briggs, 1989):

1. Let the interviewee know before the interview that you plan to take notes, and how you will use them. If a scribe has accompanied you for this reason, point that out. This provides an opportunity to diffuse the interviewee's hesitation about having his or her ideas recorded.

2. Maintain eye contact as much as possible during the notetaking intervals.

3. Don't cue the interviewee as to what is, or is not, important by furiously scribbling in response to an answer. Maintain a steady level of note taking activity throughout the interview.

4. Mark your notes to indicate transition to a new question set. For example, draw a quick horizontal line across the page to indicate a new focus.

5. If you ask a global question first, which requires the interviewee to describe a total process prior to investigating separate steps of the process, try to graph or draw it out on your notes. Later, use this graphic to help focus the interview and transition to new ideas. For example, after the interviewee has discussed the first area, point to the second "box" in the graphic and use it to help you move the interview along. While pointing to the second graphic you may say, "Now, let's examine the second area in the process."

6. Don't hide your notes—enable the interviewee to see them if it makes him or her feel more at ease.

7. Devise a code you can use to signify critical data or information of a certain type.

8. Devise a layout scheme that works for you and your project. We rarely take long, handwritten notes all across the page. More often, we divide the page up with lines. On the left, we take "normal" notes and sketch our graphics. On the right we identify heuristics, input or output that is mentioned, special performance problems associated with a point or process, documents or other reference material mentioned, or the names of people whom our interviewee thinks are expert in an area. Figure 7.7 illustrates this technique.

9. Review the notes and write the session report within 24 to 48 hours after the interview, to aid in recall.

Using Tape-Recording Devices. Taping the interview frees you make brief, high-level notes, respond effectively to the interviewee, and manage the interview process. After the session is over, you can review the tape and extract

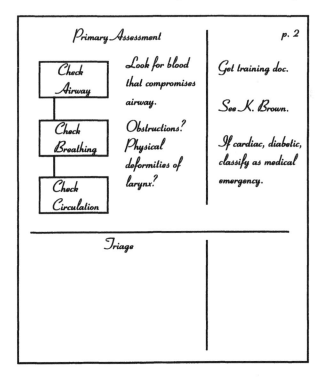

FIG. 7.7. Lay out your notes to help you capture basic material and supporting material about an area.

the important information without losing information that may be useful at another time. Furthermore, you are not faced with making a split-second decision as to what is or is not important during the interview. Additionally, tapes (particularly videotapes) enable us to compare the responses of multiple interviewees in different sessions to identify similarities and differences in their responses to an issue. Finally, we like to replay tapes for our development team or new analysts to help them familiarize themselves with the domain, if they have not had an opportunity to observe performers firsthand.

Both audiotaping and videotaping are useful. Audiotaping may be effective with a single interviewer and interviewee, but is less so if multiple participants are involved. With each additional participant it becomes increasingly difficult to tell who was speaking. If the session is a simple interview—that is, it is not combined with other techniques such as observation or protocol analysis—audiotapes may be enough. However, with audiotapes we lose the nonverbal communication that may provide clues later, as we are attempting to analyze the communication content (McGraw & Harbison-Briggs, 1989).

Both audiotaping and videotaping have disadvantages that must be addressed prior to their use. First, they add costs to the project, and require that the analyst set aside storage for the tapes (i.e., tape library) and an easy way to catalog and access them. Taping also requires that analysts address issues such as getting the right equipment, how tapes will be reviewed and transcribed (word-by-word transcription versus summary-level), and who will be responsible for the transcription.

Another disadvantage of taping is the potential negative impact on the interviewee. Some interviewees are overly concerned that what they say might reflect badly on them and be used against them in the future. Even if an interviewee has agreed to be taped beforehand, he or she may exhibit odd behaviors, such as those that follow, that should cue you to turn off the tape:

- Reluctance to commit to an answer.
- Repetition of the same response, usually from a policy and procedure guideline.
- Excessive "hemming and hawing," accompanied by furtive glances at the tape.
- The use of qualifiers that invalidate the answer just given.
- Hints that the equipment should be turned off.
- Refusal to participate.

If we observe these we typically say, "Let me stop the tape for a moment . . . ," turn it off, and see if the behavior is extinguished and if information flows in a more straightforward manner.

Taping may also encourage analyst/interviewers to become too dependent on automated record keeping. As they conduct the interview they may refrain from making any notes at all, marginal or otherwise. First, this sends a message to the interviewee that the information may not be critical. Second, it makes us vulnerable to equipment malfunctions.

These guidelines should enhance the effectiveness with which you use taping in the interview:

1. Train in the use of the equipment before you leave for your session.
2. When you schedule the interview tell the interviewee you wish to tape it and why, then ask if this is acceptable. Most people will not object if the tape will be used to help analyze the session.
3. Get there early to set up your equipment.
4. Always have extra tapes on hand.
5. Check your viewing and/or sound sensitivity range. Arrange the placement of microphones and cameras prior to the interview.
6. To reduce distraction, put microphones, cameras, and monitors in a relatively inconspicuous location.
7. If you suspect the interviewee might be reluctant to answer a question because of the tape, turn it off.
8. Always support the tape with manual notes during the session. You can use a simple outline format and summary statements. Later, you can code the notes to the counter on the tape for ease in finding key information.
9. Label the tape according to the session number and/or interviewee's name and session topic. Put the label on the tape before you leave the session.
10. Review the tape as soon after the session as possible.

Managing Interview Problems

More than any other technique, the success of the interview depends on the effectiveness of your communication and managerial skills. Problems or situations that could impact the success of the session must be addressed quickly. It is impossible to anticipate everything that could go wrong with an interview. However, we present some of the more common problems and tips for handling them in the next sections.

Getting Off Track. It is easy to let the interview get off track. A particular area may be deeper or more complex than you anticipated. The interviewee may be very interested in a topic and may wish to discuss it at great length. The interviewee may be a "storyteller" who wants to share episodes or scenarios in response to your questions. Make a continual, conscious effort to manage the interview. If you risk not completing your agenda, you must decide between continuing to focus on an area or moving ahead to the next question set. Whichever you decide, you should share your thinking with the interviewee, using statements such as:

- "That is a much broader area than I anticipated. I think we should schedule a separate session to address this area in more detail at a later time."
- "It appears you have a considerable amount of experience in this area. Could we move on to the next topic area today and schedule a follow-up session at a later date?"
- "This topic is critical to our understanding of the domain. Would it be valid to pursue it for the rest of the interview session and reschedule the other topics for a later date?"

Global Answers. Global answers are those that are so general they are difficult to translate, or have no real meaning (McGraw & Harbison-Briggs, 1989). These answers may provide useful information for you to use in structuring your general understanding of a domain. However, they are inadequate responses to the questions asked. Respond to global, surface-level answers by asking secondary, probing questions. The use of different kinds of probing questions enables you to focus the interviewee on the type and level of information required without embarrassing him or her. For example, the following clarification probes help you seek greater depth in a minimally threatening manner:

- "Say more about . . ."
- "What steps would you apply to accomplish that?"
- "Exactly how did you arrive at that?"
- "Think of me as an apprentice and describe what is involved in that."

"Yes, but . . .". This statement causes a problem because, from a communications perspective, the "but" negates everything to which the interviewee had just agreed (McGraw & Harbison-Briggs, 1989). Interviewees may use this statement in response to your paraphrase of something they just said, or in response to a statement someone else just made. With it, they dodge a difficult issue without making a firm commitment. Many people who use this technique are not even aware of it. Handling it involves focusing on the content of the statement. Try subtly rephrasing the original statement and attempt to trigger a clarified response. If that doesn't work, describe why you are confused and open the topic to a discussion of which

part the interviewee agrees with, and which part causes disagreement. You may also want to focus on a particular scenario to elicit clarification, for instance, "Give me an example of a situation that relates to this." Giving the interviewee a concrete base may help get the details you need.

Irrelevant Remarks. Irrelevant remarks are inappropriate statements. They seem to have nothing to do with the question asked or the agenda for the interview. They may even be purposeful attempts to derail the interview. Most of us are uncomfortable handling this problem because it requires that we become more assertive or aggressive. We should immediately react to the irrelevant remark using one of the following techniques:

- "Please tell me how that is related to <the topic or question>?"
- "I'm not sure I understand how that is related to <the topic or question>. Could you explain?"
- Silence—ignore the irrelevant remark and see if the interviewee clarifies or extends his or her remarks.

These statements give the interviewee the benefit of the doubt. He or she may realize that it isn't really related, or that it is only marginally related. Conversely, it *may* be related, but because you have little domain expertise, you don't see the link. In this case the interviewee can provide a more complete explanation.

Interview Interference. Other problems that can negatively impact the interview may be things over which you have little control, or things that you did not anticipate. To minimize their impact, take charge when they occur. These include:

- Noise—the environment is noisy, making it difficult to hear, concentrate, or tape the session.
- Visual distractions—activities are occurring outside the window or office area that cause the interviewer to continue to glance over and lose focus.
- Stereotyping—the interviewee makes certain assumptions about you because of your age, sex, or dress. Confidently stating who you are and explaining your assignment and expertise may counter a negative initial impression.
- Preoccupation—the interviewee is preoccupied with a device, prop, the video camera, or other item. Stop the tape or interview, note that the preoccupation seems to be distracting, and ask if something needs to done to move forward.

- Terminology—you don't understand the terminology the interviewee uses. Don't attempt to fake it. Express your lack of understanding and ask if there is a training or overview manual you could review. Usually, the interviewee will make more attempts to define or express terms and acronyms in ways you can understand if you have brought this problem into the open.

Closing the Structured Interview

The closing section of the interview enables the analyst/interviewer to review the material that has been elicited, explain how it will be used, and solicit future assistance, if needed. Unfortunately, many interviews simply end because the session ran out of time, the analyst was tired, or the interviewee left. We have analyzed hours of videotaped interview sessions and have concluded that most sessions do not have a real "closing," although it is obvious they should.

Closings are critical to the overall success of both the individual interview and future communication with the interviewee. Psychological research reveals that people tend to remember events based on the primacy–recency scheme. They tend to remember what happened *first* (primacy) in an event or interaction because their attention is at a high point. They are anticipating the event. They are attempting to predict the overall result and wondering what specific things might happen. They also tend to remember the *last* thing that happened (recency) in an event or interaction. Once again, their attention is at a high point. Their attention may have waned during the session, and certain details may have run together. They may not remember, for example, how skilled the interviewer was in extracting details. However, at the end of the session, their interest is piqued again. The interaction during this time will be seen as a part of the overall event, but will be remembered better because it is "stored at the top" of their memories.

We should make use of this tendency to show increased attention and interest during the closing. Reiterate the session goal, then use the opportunity to summarize what you have heard or learned, focusing on the main points of the interview. Ask the interviewee for any immediate clarification or points he or she would like to append or revise. (He or she may have had time to think over what was told to you and discovered that it was not completely accurate, represented a personal bias, or involved other factors that were not mentioned.) After you have reviewed interview content, tell the interviewee what will happen next, and whether you expect any further involvement on his or her part. For example, the following statement leaves no question as to next steps:

> *"I will review my notes and write up a session summary, which I would like to fax to you for your review. After I receive it back from you I will refine it before making*

it available to my project colleagues. Also, it would be helpful if you could send me a copy of the job aid that helps you code a 911 call. Here is my card."

If done well, a closing actually can enhance your rapport with the interviewee and improve your project's chance of success. The interviewees leave with a positive perception of you, your organization, your communication skills, and the session itself. For the first time, they may feel that management and a system development team actually want (and might really use) their ideas. The next time you ask for their time you are even more likely to get it. Finally, "leaving a good taste" in an interviewee's mouth increases the likelihood that he or she will mention your project favorably to others with whom you may want to interact at a later date.

Communication researchers report a variety of verbal and nonverbal techniques that can be used to close an interview. Table 7.6 provides examples of different types of closing techniques we have used successfully (McGraw & Harbison-Briggs, 1989) in structured interview sessions.

Tips for effective interview closing (McGraw & Harbison-Briggs, 1989) can be summarized as follows:

1. Even if you go slightly over time, do not cut short the closing.
2. Verbally state that the interview has come to a close, using a technique or combination of techniques with which you feel comfortable.
3. Summarize the session's major points or areas of discussion.
4. If any misunderstanding or disagreement occurred, briefly mention it and restate the resolution.
5. Provide opportunities for the interviewee to respond to a point, clarify an issue, or revise a conclusion.

TABLE 7.6
Interview Closing Techniques

Time's up	"Well, I see it is already 4:00, which is our scheduled stopping time . . ."
Declaration of close	"It looks like we have covered everything we had scheduled to discuss today . . ."
Check up	"That seems to cover the agenda thoroughly. Did we miss anything important?"
Summarization and tasking	"Good. We agree on these major points . . . We've also agreed that you will provide me with a sample reference manual by next Wednesday. I will be sending you the session notes by Friday afternoon for your review."
Showing appreciation	"Great. I feel confident that we have discussed the key points. Your preparation prior to the interview made an obvious difference."
Future plans and interactions	"Let's plan to meet on the 26th to examine the data you use to identify complications."

6. Summarize any action items or tasks that either of you is to complete. If necessary, plan briefly for the next session you will have together.

7. Let the interviewee know what will happen next.

REVIEWING AND TRANSCRIBING INTERVIEW SESSIONS

Reviewing and transcribing tapes can take twice as long as the session itself, an estimate that does not include training and setup time. This is especially true of videotapes, which provide nonverbal cues that must also be viewed and analyzed. Ten hours of audiotaped interviews with an expert planner of airlift schedules took over 100 hours to transcribe and analyze for propositional content (Hoffman, 1986). In our experience, a 1-hour interview, in which the analyst takes marginal, outline-format notes, can require at least 2 hours to review and analyze the tape after the interview. This time includes making more complete notes, rewinding to review pieces of information, coding information, or drawing graphics that depict key information.

If a word-by-word transcription is required, this process can take even longer. We favor a summary style of transcription, in which each of the topics in the agenda is addressed in a separate section. Key findings are delineated from narrative text. Heuristics are listed together with the topic area they support. Information that is questionable, or needs further clarification, is highlighted in each topic area. New terms and acronyms elicited during the session are defined to enable their addition to the domain dictionary. Documents, work artifacts, or forms that were identified and associated with each topic area are listed and attached.

After the session has been documented it will be subjected to more detailed analysis and used to provide input to domain models, functional requirements, and definitions of tasks, objects, components, and devices.

ADVANTAGES AND DISADVANTAGES OF THE INTERVIEW

The main advantage of the interview is that it is familiar to most people. Everyone has had a job interview or has solicited information from someone in formats similar to the interview. Thus, it may be more comfortable for the interviewee. It is also easy to set up, requiring little in the way of readiness activities such as selecting problems and observational environments. If managed by a competent analyst/interviewer, it can yield fairly broad coverage for a domain. Assuming the interview is structured, different interviewers can ask similar questions of multiple interviewees, and the responses can be compared or even statistically analyzed across the respondents. Finally, the interview format makes it easy to pursue an interesting area and deal with unexpected information (Kirwan & Ainsworth, 1992).

Disadvantages of the interview relate primarily to its open, communicative nature. A well-structured interview is focused and agenda-driven. It relies on the interviewer to ask good questions, make the transition effectively to new topics, dig for details, and keep the session on track. Otherwise it becomes a "gab" session that may not provide the information required, at the level of detail needed. Additionally, interviews are effective if the interviewee has stored the desired information in short term memory, or is able to retrieve it quickly from long term memory. If it cannot be retrieved or verbalized, problems result. Additionally, unlike observation sessions, which demonstrate knowledge, interviews are particularly subject to bias and may be less reliable. Finally, interview sessions are time consuming to transcribe and analyze because they are more freeform in nature than other techniques.

Defining Work Processes and Conducting Task Analysis

The analysis of work processes and tasks is a primary activity in SEP domain analysis. Through these techniques analysts are able to investigate tasks the system will support, define information and data needs, determine the level of intelligent support required, and plan the functionality of the components.

Work process and task analysis are two levels of the same activity. The goal of work process analysis is to identify and define a performer's major job functions, as well as to enable us to use them to:

- Understand quickly the areas our new system must support.
- Identify initial opportunities for re-engineering.
- Bound the scope of our task analysis.
- Guide us in completing task analyses.

Work process analysis yields top level information, whereas task analysis enables the decomposition of each work process into its component tasks. These tasks are then analyzed to determine task goals, operations, and requirements. Work process analysis results in a set of flowcharts or scenarios that depict activity at a top level, which becomes the input for a task analysis. The output from a task analysis becomes the task model that describes tasks and subtasks that support specific functions or performance goals, and in some cases (i.e., for training systems), the skills, knowledge, and abilities on which effective task performance depends.

Understanding task-related issues can assist developers in several ways. First, it enables the developer to take a user-centered design approach. If we understand the performer's work processes and tasks, we can define system requirements, functionality, and user interfaces that will make it easier for the performer to complete his or her job. In addition, understanding the performer's tasks also helps us recognize opportunities to improve performance by re-engineering existing tasks. Re-engineering may be as radical as identifying a set of tasks that may be eliminated altogether, or identifying a system design that can eliminate or automate some subtasks. Alternatively, it may be as simple as designing systems that provide concurrent distribution of information to alter the previously linear nature of the task (e.g., an approval cycle that requires multiple signatures).

The purpose of this chapter is to define and describe work process and task analysis within the context of SEP domain analysis. We first present background information on work process analysis, including some of its variations, followed by guidelines for conducting a work process analysis. Next, we define how task analysis can be used for system development, including basic definitions and requirements. Finally, we present guidelines for conducting major types of task analysis, present examples of task analysis output, and identify advantages and disadvantages of the techniques.

USING WORK PROCESS AND TASK ANALYSIS

To conduct an adequate domain analysis and define preliminary requirements, analysts must identify the primary set of work processes or job functions that the target performers are expected to complete. These are the top level accomplishments, as opposed to specific, individual tasks. To help identify processes or job functions, analysts should examine the outputs a performer is expected to produce. For example, a nonprofit funding entity might have a work process such as *Select proposals for funding*. This is a high level job function that the performer must accomplish. It will result in a set of proposals recommended for funding. It comprises many tasks that must be completed to produce the required output.

To understand a work process well enough to decompose it into its individual tasks, analysts should identify the following (McGraw, 1994c; McGraw, 1994e):

- *Job responsibilities*—what do performers have to accomplish? For example, "Filter solicited and unsolicited proposals and recommend those whose goals parallel organizational goals" is a responsibility. It specifies what the performer must do.
- *Source of the work*—from where or whom does the work originate? For example, there are multiple sources for the work process "Select pro-

posals for funding." One source is external, including graduate and undergraduate grant proposals and proposals from nonprofit organizations. Another is internal, in that corporate executives may sponsor a proposal that came in directly to them. These two sources may trigger slight variations in how the work process is completed.

- *Types of output and output destinations*—what is produced when the work process is complete, and what is the destination for the work-products? For example, when the process "Select proposals for funding" is complete, the result is a list of recommended proposals and their suggested funding amounts. The list will be received and acted on by corporate executives at the VP level or above.

- *Critical success factors*—what determines whether performers are successful in accomplishing this work process or job function? Examples of critical success factors include time (proposals must be rejected or recommended within 30 days), accuracy (90% of the proposals suggested for funding should be acceptable to upper management), and so on.

- *Means of measurement*—how is performance measured to determine if the critical success factors are met? Some work processes are easier to measure by quantitative means such as time to complete review, time spent in review, and number of proposals processed. Others may be more amenable to qualitative measures, such as accuracy and upper management acceptance.

- *Volume or impact on total job*—what percentage of time is spent performing this work process? For example, one set of performers may be focused 100% on a single work process such as "Select proposals for funding." Others may be responsible for additional work processes, and may only spend 50% of their time selecting proposals for funding.

- *Motivational issues*—what motivates or demotivates performers to accomplish the work? Do they receive a bonus for recommending a proposal set that is unanimously accepted, are they compensated based on the number of proposals they recommend, or are they rewarded for rejecting the highest number of proposals?

IDENTIFYING AND ANALYZING WORK PROCESSES

Developing a system that gives its users what they need requires a solid match between the performer's work processes, including its goals and work flow, and their expectations about how the system will function. The best way to suit a system to its users is to base it on their mental models (McGraw 1994d; McGraw, 1994e). Such a model accurately depicts what users do and how they perform their work. Detailed knowledge about the structure and flow of the work process and its tasks allows designers to structure systems to support and extend the work (Holtzblatt & Jones, 1993). We obtain this detailed knowledge by eliciting and analyzing work processes and tasks within the performer's environment.

Using Observation, Scenarios, and Interviews to Identify Work Processes

In chapter 6 we defined how observation is used to understand the total job for which a performer is responsible. Conducting observation sessions helps you visualize the work processes and tasks you will be analyzing. The real-life context in which work is performed is the best environment for identifying work processes. During these observations you collect scenarios that describe real situations that occurred, problems that were solved, decisions that had to be made, and work that was done. These real-life narratives provide fodder for understanding the work processes. Analyze selected scenarios for your target performers to identify the major work processes that are accomplished in the scope of each scenario.

For example, early sessions with target performers in the trauma care domain yielded dozens of scenarios like the one shown in Fig. 8.1. Each item in bold text represents a major work process. These processes can be used as stimuli for the task analysis.

The major work processes that can be identified from the scenario are presented in Table 8.1. Each of these work processes, in turn, could be decomposed into the specific tasks that must be completed for the work process to be accomplished. For example, "Conduct triage" involves the following tasks:

- Determine mechanism of injury.
- Determine extent of injury.
- Determine initial vitals.
- Code victims as Immediate, Delayed, Minimal, or Expectant.

In addition to observations, analysts can use interviews to define or, better yet, refine work processes. Most performers are comfortable, and even enjoy talking about their major job responsibilities—what their job entails and the major objectives they must accomplish. Consequently, an interview with selected performers, either individually or in small groups, may yield enough information to lay the groundwork for a work process analysis.

However, analysts must realize the limitations of using only an interview to identify work processes. Chapter 7 provides guidelines on when an interview is effective and how to conduct good interview sessions. Remember that interviews primarily provide declarative knowledge. Without interactive

During a rush-hour ride home on Dorsey Road (a well-traveled 4-lane road with no median strip), a middle-aged woman swerved to miss a ladder lying in the road. Moving westbound at 45 mph, she crossed the center strip, where she was hit by an eastbound car. The impact spun her car back into the westbound lanes, caused the door of her car to open, and threw her onto the roadway, where she was struck by another westbound car. In an attempt to avoid hitting the woman's car or her, two other westbound cars collided. 911 was called by a traveler, and the appropriate **dispatch center managed the initial response** to this 4-car collision.

First on the scene were the police and fire units. They **assessed the scene** and determined that extrication equipment and a medevac helicopter would be needed and dispatched this request. They **cordoned off** the section of the road to ensure the safety of the EMS providers and the victims.

When the EMS units arrived they began their **triage of the victims** and each unit was assigned victims to manage. The triage included sorting victims into categories based on their type and extent of injury, and on the likelihood of their survival.

An assessment was performed on each victim, including a primary assessment to determine the victims' airway, breathing, and circulation (i.e., ABCs), and a secondary assessment. One victim, who was trapped in his car, required that the medic enter the car to assess the victim and begin initial lifesaving treatment. As the extrication equipment arrived, it was used to free the trapped victim whose assessment was then completed.

Next, **on-scene treatment** began, in response to immediately life threatening situations. EMS personnel began making transport arrangements and contacting receiving hospitals for medical support, if needed. As each patient was stabilized or deemed ready to transport, he or she was loaded onto the ambulance unit or helicopter, depending on severity of conditions, and **transported to the appropriate receiving facility**. During transport, treatment was continued. Upon arrival at the receiving hospital, the victim, and information about him or her, was **transferred to hospital personnel**.

FIG. 8.1. Sample scenario highlighting emergency response work process.

TABLE 8.1
Sample Work Processes for Emergency Response

Work Process	Performer	Output
Dispatch call	911 Dispatch operator	Event, coded according to priority, and selected units for response
Assess scene	First responder Emergency medical personnel	Approval to enter scene and provide treatment
Conduct triage	Emergency medical personnel	List of victims according to criticality and treatment order
Assess victims	Emergency medical personnel	Initial vitals Initial injury description
Provide on-scene treatment	Emergency medical personnel	Stabilized victim
Transport victims	Emergency medical personnel Transport drivers/pilots	Stabilized victim delivered to appropriate treatment facility
Transfer to receiving facility	Emergency medical personnel Treatment facility personnel	Victim admitted to treatment facility

observation we risk getting simply the performer's summary or high-level description of the work process. It may be what the performer recalls, but it is likely not complete or accurate. Therefore, we urge analysts and designers to combine the interview with an observation to enhance its effectiveness, elicit deeper knowledge, and develop a more complete understanding of the work processes.

Regardless of the type of interaction you have with performers, we suggest that you submit a presession worksheet to the performers before you meet with them. A presession worksheet has the following benefits:

- Enhances the quality of information performers provide.
- Diffuses performer anxiety about the session.
- Enhances your preparation prior to the session.
- Enables performers other than those you interview to feel as though they have had input to the process.

First, as they review the presession worksheet, performers think about their work processes, the percentage of time spent on each work process, output for each work process, and critical success factors related to each work process. Because some of this information may be stored in long-term memory, it gives them an opportunity to identify and retrieve important information in a nonstressful situation. As a result the analyst gets better, more thought-out answers.

Second, the performers have time to think beyond their first impressions and responses to the worksheet. They know that they have the information, but their first response may not be complete or accurate. However, because they complete the worksheet prior to the session, they have opportunities to change their answers as they retrieve information, gather data, and respond. Thus, when they come to the session they are more likely to be comfortable, and confident of their information.

Third, a presession worksheet also enables you to review potential interviewee's responses prior to conducting an interview. You may decide that you want to meet with certain respondents because their information was clear and detailed, but that you do not need to meet with others. It also enables you to review all worksheets prior to conducting sessions and to become familiar with the terminology and work processes you expect to encounter.

Figure 8.2 illustrates a sample presession worksheet that was used in the public affairs department of a major corporation[1] to help identify future systems that could re-engineer the way work was currently being accomplished. This worksheet was sent to 20 potential interviewees prior to meeting them. It was accompanied by a one-page description of the purpose of the

---

[1]This worksheet was developed in 1993 by Karen McGraw of Cognitive Technologies, and modified by Kate Blodgett of RWD Technologies, Inc.

Work Process	% of weekly activity	Systems or artifacts used	Interactions with others required?	Typical problems completing
Accepting and logging proposals	40%	WIKET MIRE Form 1345	Yes— S. Jones M. Miller R. Smithson Proposal author	Too many steps Redundant entry of information Time lags System reliability

FIG. 8.2. Partial response to a presession worksheet to identify major work processes for a particular department prior to follow-up interviews.

project and directions for faxing it to the analysts. It was used to prepare the analysts for the sessions and to help select the performers with whom we would conduct an interactive observation versus a telephone interview.

Finally, a presession worksheet can be used as a political tool. Few projects have the luxury or need of interviewing each performer and supervisor. However, some politically sensitive projects or departments may need to demonstrate that multiple parties have had input. The presession worksheet provides a vehicle for soliciting general information from many people.

Using Document Analysis to Identify or Define Work Processes

Some domains have extensive training material and other support documentation that you can use initially to identify and define preliminary work processes. Later, you should confirm and modify these by interacting with performers during observations and interviews.

Document analysis can help you:

- Identify the primary work processes.
- Analyze the relationship of work processes to each other.
- Define the work process elements that help you understand it better (e.g., identify critical success factors, heuristics used, goals).

You may wish to use a tabular format to help you capture and organize work process detail about a document. Figure 8.3 shows an example of a layout for a work process detail matrix for document analysis.

Keys to Successful Identification of Work Processes

Work process analysis is not hard, nor is it as tedious as task analysis. However, a few factors do impact the success with which you accomplish it.

Source Document & Page Reference	Work Process	Stated Goal	Critical Success Factors	Related To

FIG. 8.3. A table format can be used to help organize document analysis for work process identification.

Identifying the Right Performers

Current trends in systems development are highly user- or performer-centered and emphasize participatory design (McGraw 1994c; Schuler & Namioka, 1993). That is, systems are designed to meet the needs of the workers. All systems must meet the "what's in it for me" (WIIFM) user test (Cjelli, 1994). Potential system users must be able to see a need for the system and understand how it can assist them in accomplishing their work processes better, faster, or with greater ease. Meeting the WIIFM test requires that the potential users participate in all phases of the development process.

This begins with the capture and analysis of scenarios. Real performers and potential system users provide scenarios that reflect actual episodes. We observe performers in the context of their jobs. We work with actual performers to help identify the work processes that are reflected in the scenarios. We work with performers in focus groups across departments, divisions, or job functions the new system will affect, to ensure that we understand the total work processes (McGraw, 1994c). Granted, we may need to meet with supervisors and managers to better understand the total work process, but the primary activity is focused on the performer.

Any of the following will make it difficult to meet the WIIFM test and to ensure that information about the work process is accurate:

- The politics of the organization inhibit your ability to talk to the real performers or potential users.
- One person, usually a supervisor, acts as gatekeeper and demands that he or she serve as the sole interviewee because he or she has a better, more complete picture of the process. (After all, it is normally part of a supervisor's job to "protect" the workers and ensure that work gets done.)
- You are assigned marginal performers or those that are "not busy" with whom to conduct the work process analysis.
- One group is left out because the system will change or do away with their jobs. Obviously, to develop a system that has complete functionality we must understand how the processes are conducted today and where the major potential for productive change exists.
- Performers are "allowed," but not motivated, to participate. For example, their work goals are not relaxed during the time they meet

with you, which means that by spending the time necessary to assist you with a work process analysis, they are not meeting their performance goals (McGraw, 1994b).

Identifying and Analyzing Key Documents

As previously discussed, it is best to analyze actual work processes in the context of the performer's work environment. However, if you are working in a new domain, or if the domain is very complex, you may find it useful to conduct an initial work process analysis by reviewing key documents. Candidates for such an analysis include new hire training material, job aid documents and similar work artifacts, special training topics, or materials that prepare performers to meet certification requirements.

Be sure to verify (through observation, interactive interviews, or focus groups) information obtained via document analysis, including any initial diagrams or work process details. Work is not always performed as it is prescribed or taught.

Representation Techniques for Work Processes

It is critical that you extract and identify the work processes from within the context in which they occur—through scenarios or observation—and that you communicate your findings effectively to other development team members. Part of this communication is textual. This includes selecting accurate descriptors of the work processes or job functions. It also includes using tables and matrices that organize the work processes in a way that can be understood easily (e.g., read, compared, discussed). Another part of this communication includes graphic structures, such as flow charts, that depict the flow or structure of the work. Development teams should decide at the beginning of the program in which formats they wish to represent work processes, select tools that support these representations, and develop table and graphic templates to use as starters and models. For example, Fig. 8.4 depicts the top level work process of a team of designers at an advertising agency.

Reviewing the Work Process Analysis

After you complete the work process analysis, you will have a table or flowchart that depicts the elements you wish to analyze. This includes outputs, performers, resources used, critical success factors, volume, input sources, and destination of work process output. Review the results of your work process analysis to:

- *Identify performers with whom more interaction is needed*—some performers may be more able to provide information relevant to specific work processes and should be selected specifically to meet session goals.

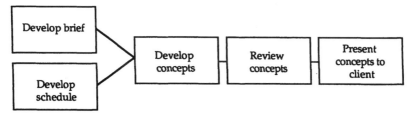

FIG. 8.4. Ad agency top level work process diagram, representing work process that takes from two to six weeks.

- *Identify areas of criticality*—it may be evident that some critical success factors and goals are more important than others. This information can be used to divide session time appropriately to manage the risks inherent in not getting enough information in these important areas.
- *Identify risks*—the work process analysis may have revealed numerous risks, such as how political a topic or area is, that resistance to change may be an issue, or that incentives are dysfunctional. This information can be used to prepare for further discussion and to ensure that all parties are aligned on the purposes of the project.
- *Prepare for the task analysis sessions*—the output of the work process analysis helps you become acquainted with the accomplishments the performers attempt to attain, the terminology they use, and the factors that determine success. Task analysis should only be undertaken *after* you understand the work processes that provide the context for each task.

TASK ANALYSIS: DECOMPOSING WORK PROCESSES

Task analysis is a methodological tool that can be used to describe the functions a human performs and that determines the relation of each task to the job function or work process (Shannon, 1980). It was originally used in the training domain, but has since been used in numerous other domains.

We use the set of work processes as the input to the task analysis. We examine each work process to determine if it is a target for task analysis. For example, if a particular work process is outside of the scope for a system we are developing, it will not be targeted for decomposition and further analysis.

Next, we use task analysis to decompose each selected work process and determine what steps or tasks must be completed to accomplish it. The level of detail to which we go should be determined by the needs of our system development efforts (see *Planning for task analysis*).

What Is a Task?

Task analysis is difficult to teach and do because of the complexity of the concepts involved. There is great variance in the definition of *task analysis*, and in descriptions of how it should be performed. This variance is evident even in the definition of *task*. A task is generally considered to be "a set of activities, co-occurring at the same time, sharing some common purpose that is recognized by the task performer" (Miller, 1967, p. 11). A more narrow view defines a task in terms of transfer of information between components of a person-machine system with an emphasis on the operations performed on the information within a component (Teichner & Olson, 1971).

Hackman (1968) offered an even more explicit definition of a task as a set of activities "assigned to a person by an external agent or is self-generated, and consists of a stimulus complex and a set of instructions which specify what is to be done. The instructions indicate what operations are to be performed by the performer with respect to the stimuli and/or what goal is to be achieved" (p. 12).

Fleishman and Quaintance (1984) noted that these differences in definitions occur because *task* is a conceptual construct. Constructs used to differentiate human tasks require that we make deductions and inferences about the dimensions of a task. Our challenge is to identify the task conceptualizations and characteristics most useful in enabling us to "derive meaningful predictions and generalizations about human performance" (p. 50). Furthermore, the information we capture about tasks must be in a format that is useful in defining system functionality, component design, and human–computer interfaces.

Anatomy of a Task Analysis

The purpose of our task analysis is to investigate scenarios and observe work processes so we are able to classify and decompose tasks involved in carrying out a particular job function or performance goal. After tasks and their subtasks have been identified, we can define the skills, knowledge, abilities, information, and materials required to complete each task set. To provide this type of information, a task analysis must include specific pieces of detail about each task. Information that may comprise a task analysis is depicted in Fig. 8.5 and Fig. 8.6.

The following list defines some of the more common elements of a task analysis:

To describe this information	Use elements such as:
Description	Text description of task Task goal statement
Requirements for undertaking	Initiating cues or event Personnel needed Prerequisite training
Task area environment	Job location
Nature of the task	Actions required Decisions required Discrete vs. sequential Complexity, difficulty Communication required Task criticality Attention required Memory load requirements

FIG. 8.5. General task information often included in task analyses.

To describe this information	Use elements such as:
Performance on the task	Required accuracy Required speed
Inputs	Input, information used Information source
Outputs	Output or feedback Destination of output
Consequences/problems	Typical errors Error consequences Hazards encountered
Decision making factors	If X, then Y Constraints
Tools used	Systems used Equipment, devices used Work artifacts used

FIG. 8.6. Specific task information.

- Primary task name—a short, descriptive title describing a function.
- Task number—indexing number for easy reference (i.e., 1.0, 1.1, 1.2, etc.).
- Task description—general information that defines the task, including actions, aids, and feedback.
- Task type—identification of the task by one or more categories, such as perception, mediation, communication, and motor (see Table 8.6) to assist in comparing it to other tasks, or to understand the predominance of one or more task categories in a work process.
- Associated subtasks—specific steps or smaller tasks that must be completed to accomplish the task (i.e., further decomposition of the task).
- Knowledge or information requirements—types/levels of knowledge or specific information required to complete the task.
- Skills required—prerequisite skills that are required for task completion.
- Standards—standards that must be met for the task to be performed correctly and completely.
- Enabling activities—tasks or actions that must have already been completed.
- Personnel required—who performs, or is qualified to perform, the task.
- Performance time—typical completion time or critical response time.
- Related tasks—tasks related to, depending on, or interacting with current task performance.
- Special cues, checkpoints, and values associated with a task.

Note that not all of these are required in any one task analysis. In task analysis, final form depends on the function and use to which the information will be put (McGraw & Riner, 1987). This information can be used to develop or refine requirements, identify object attributes and actions, and identify the need for training or performance support.

Planning for Task Analysis

The following guidelines should help you plan successful task analysis efforts:

- Agree on the purpose and depth of the task analysis.
- Determine the desired output of the task analysis.
- Determine how tasks will be categorized.
- Provide training in task analysis.

The sections that follow provide information that can help you define the task analysis activity for a given project.

Determining the Purpose and Depth of the Task Analysis

First, define the bounds for the task analysis by clearly specifying its goals and purpose. Many analyses have failed because management did not understand how detailed the task analysis needed to be. Not all work processes must be analyzed in further detail if they will not be addressed by system functionality. Furthermore, some tasks may already be well understood and need no further analysis. Is the purpose of the task analysis to define embedded performance support components that automate the work, or simply to design performer-centered screens? The time and depth of analysis required differs according to the answer to this question.

Analysts are cautioned to carefully consider how much task detail they need in order to meet project goals. Do not spend more time than is required. We have found that no more than two or three task levels are necessary to understand a task well enough to determine system requirements and functionality.

Determining the Output of the Task Analysis

The development team should determine early in the project the desired output forms or representation of task information. What will the finished output of the task analysis be, and in what format will it be communicated? Define templates prior to the analysis to ensure that the right type of information is captured at the right level of detail. Data should be collected using these agreed-on forms to reduce time-consuming translation into other formats upon completion of the analysis.

Categorizing and Naming Tasks

The development team must also determine how tasks will be categorized and the vocabulary that will be used to describe tasks. For example, if two team members use the term "Monitor . . ." in a task statement, the type of task represented by the statement, and the activity being performed, should be relatively similar. Without this agreement, there will be very low interrater reliability—that is, two analysts might observe the same task, but name it differently.

One of the first task features that should be determined is the task's nature. Task nature can be compared in a number of ways. First, tasks can be categorized based on the type of knowledge they require. A task such as "recognizing objects" or "identifying object labels" may require only syntactic, or surface-level knowledge.[2] However, applying rules to arrive at a decision requires more semantic, or deeper knowledge. To "make a proper

---

[2] We discuss knowledge types in chapter 4.

diagnosis" in a troubleshooting domain, an expert may rely on a combination of semantic knowledge and episodic, or case-based, knowledge.

Another way to categorize tasks is to discover whether the task is discrete (i.e., stands alone) or sequential (i.e., tightly linked to the task that precedes or follows it) in nature. Although people find it relatively easy to determine which task or subtask comes next, it is difficult to decompose, think about, and discuss each task as a separate unit. The response pattern for discrete tasks is such that one step or task does not invariably lead to another specific task. Furthermore, time lags are more likely to exist in discrete tasks, whereas sequential tasks are usually tightly linked to one another. These differences increase the complexity of the task because it is more difficult to predict specific performance paths. Figure 8.7 illustrates the difference.

We analyze task nature early in the task analysis process because it is a key feature in helping determine the possible strategies a performer applies to complete a task (Hogarth, 1974). These strategies may help us better understand the type of knowledge (i.e., syntactic, semantic, episodic, procedural) that is being applied, the complexity of the task set, and the technique that will be most effective in analyzing the tasks. For example, tasks that are sequential and represent either procedural knowledge or episodic knowledge may be analyzed effectively using techniques such as timeline analysis, observation, or operational sequence analysis. However, these same tasks would be nearly impossible to trap and analyze using the interview.

Regardless of the type of knowledge tapped during task completion, the nature of the tasks, or the selected technique, the developer must construct an effective system for classifying and describing tasks. Unfortunately, this has been a source of difficulty because two analysts may use different terms or structures in describing the same task. This makes it difficult to recognize

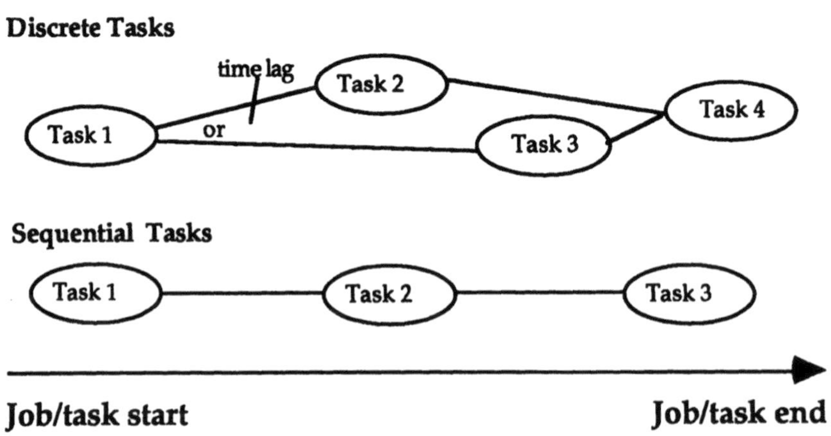

Discrete Tasks

Sequential Tasks

Job/task start   **Job/task end**

FIG. 8.7. Discrete versus sequential task completion.

redundant tasks in a set compiled by multiple analysts, or to communicate clearly in system documentation (e.g., requirements documents, design documents, etc.). It is helpful to have group brainstorming sessions with performers and other analysts to refine initial task lists.

Prior to selecting the technique and meeting with performers, we strongly urge developers to define the form of the resulting task analysis document and how it will be used, and to identify essential vocabulary and structure that can be used to describe tasks. Describing a program-wide set of acceptable behavioral classifications and task verbs enables a team's developers to understand each other's output. It also ensures that other people can study and use the information later.

Task Classifications. Numerous researchers have defined various classification schemes for tasks. Most of these schemes are descriptive in nature; however, as some authors note (Fleishman & Quaintance, 1984), they do offer some guidance for ordering or arranging task entities into groups on the basis of their relationships. Few of these classification systems go to the extent of defining precisely what specific task terms mean. Therefore, developers should investigate classification schemes and select one that they can modify and use to define task descriptors. A classification system should meet the following criteria:

1. Task categories and descriptors should be defined as precisely and objectively as possible.
2. The development team should be able to apply task descriptors fairly reliably so that two analysts observing the same term would categorize and name it in a similar fashion.
3. The classification and naming system devised should fit the domain. That is, it should be easy to observe the task and select a name and category for it.
4. The classification and naming system should meet the goals of the project. For example, the set of descriptors need only be large enough to document the tasks that are relevant for the system development project.

Categories used to describe tasks are perception, mediation, communication, and motor (Woodson, 1981).[3] Perception tasks are characterized by recognizing and identifying; examples include Scan and Detect. Mediation tasks are characterized by actions on information that change its format or representation; examples include Categorize and Compare. Communica-

---

[3]This scheme was originally known as the Berlinger Classificatory Scheme (Berlinger, Angell, & Shearer, 1964).

tion tasks are characterized as involving tasks common to the transmission of information; examples include Advise and Request. Motor tasks are kinesthetic in nature or require tactile feedback; examples include Move and Press.

We have used this classification scheme on a number of projects and have found it useful and fairly complete (McGraw & Harbison-Briggs, 1989). Furthermore, it has been well-researched and tested in a number of domains. Christensen and Mills (1976) found this system effective in determining what operators do in complex systems interactions. They used these classifications to describe the activities of systems operators in psychological terms, and to provide estimates of the proportion of time spent on each activity. They found satisfactory agreement between two independent raters who classified the activities of nine operators with different jobs.

Assuming that the characterizations in this scheme are sufficient, an analyst team could enhance its use by following these suggestions:

1. Review standard terms or task descriptors for each type of task, then review scenarios that have been generated and catalogued for use in the domain. By reviewing these scenarios and comparing task activities to standard terms, the team can develop an initial list of terms that could become the "standard set" for all task analysis activities in their particular domain.

2. Define all performer activities in clear, unequivocal terms.

3. Define performer activities in sufficient detail to discriminate between tasks.

4. Involve performers in defining task descriptors and/or in rating tasks to ensure accuracy and completeness.

Tables 8.2 through 8.5 detail terms that are considered by task analysis experts to represent perception, mediation, communication, and motor tasks.

As you can see, no one domain will require extensive use of task terms in all categories. For example, Table 8.6 lists standard task terms in each category for telephone agents in a particular type of call center. Note the predominance of communication and mediation terms, and the scarcity of motor terms, because of the domain.

TABLE 8.2
Sample Perception Task Verbs

Detect	Inspect	Observe
Read	Receive	Scan
Survey	Discriminate	Identify
Locate		

TABLE 8.3
Sample Mediation Task Verbs

Categorize	Calculate	Encode
Compute	Interpolate	List
Tabulate	Translate	Analyze
Select	Compare	Estimate
Predict		

TABLE 8.4
Sample Communication Task Verbs

Advise	Answer	Communicate
Direct	Indicate	Inform
Instruct	Request/Ask	Transmit/Send

TABLE 8.5
Sample Motor Task Verbs

Activate	Close	Connect
Disconnect	Join	Move
Press	Set	Raise/Lift
Regulate	Hold	Lower
Adjust	Track	Align
Synchronize		

Conducting the Analysis

To complete a task analysis, first detail the complete set of task statements that describe what an individual performing the specific job function should be able to do. This begins with identifying the following:

- Goals, responsibilities, and key accomplishments of individuals performing the job.
- Normal conditions under which the task is performed, including time requirements, volume, and current use of tools, job aids, or other handbooks and materials.
- Tasks requiring input from other sources or individuals, and those resulting in output to another destination or individual.
- Tasks with embedded decision points or problem solving activities and the source for heuristics and rules used during this process.
- Which tasks in a performer's job will be analyzed and, if required, sample case material or simulated tasks to use during this activity.

TABLE 8.6

Sample "Standard Set" of Terms From a Call Center Domain, Reflecting the Impact of Domain on Task Categorization

Perception	Mediation	Communication	Motor
Detect (i.e., new call)	Compute (i.e., rates)	Advise (i.e., on best service type)	Activate (i.e., computer system)
Scan (i.e., account information)	Select (i.e., screens)	Direct (i.e., customer to manned location)	Disconnect (i.e., phone and computer)
Receive (i.e., call)	Predict (i.e., customer needs)	Instruct (i.e., how to fill out airbill)	Press (i.e., keyboard keys and mouse button)
Discriminate (i.e., scam vs. authentic call)	Interpolate (i.e., actual needs from pieces of information)	Answer (i.e., customer questions)	Adjust (i.e., telephone volume; monitor brightness)
Read (i.e., online documentation)	Translate (i.e., lbs into kgs.)	Indicate (i.e., which airbill fields must be complete)	
Identify (i.e., caller requests)	Compare (i.e., benefits of 2 services)	Request/Ask (i.e., questions to gather information)	
Locate (i.e., pickup detail)	Encode (i.e., rules into words the customer understands)	Communicate (i.e., findings with customer)	
Detect (i.e., close time field on screen)	List (i.e., types of service)	Inform (i.e., customer of package loss)	
	Analyze (i.e., customer information to identify requirements)	Transmit/Send (i.e., fax directions to customer)	
	Estimate (i.e., transit time)		

In addition to these guidelines, the following will assist you in developing usable task analyses:

1. Define a statement of purpose to help provide structure and to aid in selecting the appropriate elicitation and representation technique.
2. Obtain a thorough knowledge of the environment, the current "system" in place, and the specific function or job under examination. Use expert opinions, on-site observations, literature, work artifacts (e.g., forms, sample reports, etc.) and existing task descriptions.
3. If appropriate, use a matrix technique to define the tasks. For example, define a job by its functions; then define tasks matrixed to the functions. Next, define subtasks, knowledge, skills, objects, or attributes associated with each task. Matrices enable auditability and help ensure complete coverage of a target.

The success of the task analysis depends on the quality of the basic unit, the task statement. According to Mager (1962), task statements must consist of the behavior, conditions for the behavior, and the standards against which the behavior is evaluated. The following guidelines are useful in ensuring the appropriateness of a task statement:

- Use of an action verb, like those presented in the previous tables.
- Existence of only one behavior per task statement.
- Ability to observe or measure the behavior that is described.

The behavior portion of the statement should clearly indicate a single operation or activity. The conditions portion of the statement should describe the performance environment and the resources a performer uses to complete the task. Task performance standards should include minimum acceptable performance levels, time limits, and quality/quantity levels, if known. The following guidelines were established for constructing task statements during the development of knowledge-base systems (McGraw & Riner, 1987; Riner, 1982):

1. Each task has a definite beginning and end. It is performed in a relatively short period of time.
2. A task statement describes a finite, independent part of the job.
3. Each task is mutually exclusive, although it may be sequentially linked to another task.
4. Each task statement should be clearly understood by another performer doing the job. Its description should use terminology that is consistent with current usage by the performer group.
5. Each task should be "ratable" in terms of the amount of time spent completing it.

6. A task is begun by an observable cue.

7. A task statement is brief.

8. A task statement avoids qualifiers. Only when there is more than one way to perform a task is a qualifier used.

9. A task statement uses one verb, except when several actions are invariably performed together.

10. A task must be measurable. The end result or product can be estimated or measured as having been done correctly.

USING DECOMPOSITION TECHNIQUES TO ANALYZE AND PREPARE OUTPUT

Various types of decomposition techniques are effective in representing the analysis and preparing output to be shared with other team members and customers. The following set reviews operational sequence analysis diagrams, hierarchical task analysis, and cognitive task analysis. Whenever feasible, sample representations or output samples are provided.

Table 8.7 compares decomposition techniques you may use to analyze tasks and prepare output. Tips for using and preparing each of these techniques is described in the sections that follow.

Hierarchical Task/Subtask Analysis

Hierarchical task/subtask analysis is one of the most commonly used techniques for analyzing and representing events or steps within a process. It helps the analyst decompose and analyze the actions of a performer. At the very least, it requires that the analyst describe a task in terms of the operations (i.e., what we do to attain goals) a performer completes to attain a goal. It may also depict the plans (i.e., conditions when a set of operations should be carried out) a performer should follow in attaining that goal (Kirwan & Ainsworth, 1992). Hierarchical task analyses can be presented in tabular form, as a graphic, or in task frames.

It is essential that the task analysis identify the separate tasks involved in an activity, describe it succinctly, label it for later reference, and provide any other information important for the project. It is not enough to know that one step in activating an emergency response is to "Identify emergency situation." It is more useful to understand that this one task involves the successful completion of the following subtasks:

- Recognize buzzer/alert.
- Check gauge for temperature =202+.
- Confirm temperature emergency.

TABLE 8.7
Decomposition Techniques for Task Analysis

Decomposition Technique	Description	Purpose
Hierarchical Task Analysis	Tabular format depicts task analysis information in a table format. Hierarchical analysis format depicts task decomposition graphically. Task template format depicts a summary level view of a task.	Use to explicate the steps a performer takes to complete a task.
Operational Sequence Analysis	Diagram format is a flow chart that categorizes operations into behavioral elements and represents them with standard symbols. Temporal diagram depicts tasks that are highly sequential in nature.	Use to investigate the operations associated with a task, and the order in which they are performed, and to identify tasks performed by people and those completed by computer systems.
Cognitive Task Analysis	Tabular format depicts tasks and various cognitive factors that can be rated for each task. Results can be statistically analyzed.	Use to identify tasks that are mentally challenging, difficult, require memorization, have high consequence of error, or present critical performance problems.

Purpose

Hierarchical task analysis depicts the decomposition of a task using text tables or hierarchical task/subtask diagrams. The form it takes may vary, depending on the goal (i.e., "form follows function"; McGraw & Harbison-Briggs, 1989). Regardless of the form, the purpose is the same—to explicate the steps or operations a performer undertakes in the completion of a task, and to depict additional steps if needed.

Tabular Hierarchical Task Analysis. Tabular hierarchical task analysis simply organizes task analysis information in a table format. This technique is very efficient because the information it presents can be restyled into other formats, including a graphic hierarchical task analysis. This method depicts the primary goal, first level tasks or operations, and subtasks that comprise each task. *Plans* indicate methods used to attain goals. *Tasks* are the operations actually completed to attain the goal. *Subtasks* are decompositions of the task at levels that may need to be known. Figure 8.8 is an example of a table format depicting a hierarchical task/subtask analysis.

Plan/Goal	Task or Operation	Subtask
Ensure scene is safe for emergency personnel to work and determine the need for additional assistance.	1.0 Conduct Scene Assessment	1.1 Identify environmental hazards in the area (downed wires, hazardous materials, etc.)
		1.2 Determine general safety of area (e.g., gangs, guns, etc.) based on previous cases and type of incident
		1.3 Estimate accurate number of victims and type of incident (e.g., gunshot wound vs. standard medical problem)
		1.4 Determine if additional assistance (e.g., police) is needed to ensure safety

FIG. 8.8. Partial tabular hierarchical task/subtask analysis.

We can expand this representation to include other information about the tasks or subtasks, such as:

- Personnel/other performers involved.
- Information used.
- Decisions associated with a task or subtask.
- Constraints or problems.

Figure 8.9 is an example of a task/subtask matrix, which is an expanded tabular representation of the steps a performer takes to complete a task. We have used it to enhance communications between analysts, developers, and performers on various SEP projects.

Hierarchical task analysis can also be represented graphically. For some analysts this format is a natural choice because they may already record notes in a graphical form during a session. Figure 8.10 illustrates a graphical hierarchical task analysis representation.

Task Frames. Another representation is the task frame, which is useful because it depicts a summary level of a task. After we have conducted a task analysis and identified the elements already discussed, including task name, task number, description, subtasks, information, personnel, equipment used, etc., we compile this information using a template constructed

Task Element	Purpose	Decision	Action	Information Required
1.0 Conduct Scene Assessment	Ensure you can safely treat casualties	Determine if scene is safe for emergency medical personnel Determine probability of continuing or escalating hazard	Visual scan (Constraint: Must not put medical resource in extreme danger) Count victims	Nature of the agent that caused the trauma Number of victims Threat level around the scene
2.0 Conduct Casualty Assessment	Determine who needs care most immediately, and what type of care		For every casualty, do 2.1, then 2.2 and 2.3	
2.1 Conduct triage	Determine who needs care most immediately	In non-threat environment, save person who is the most critical In threat environment, get as many back to the line as a fighting resource as possible Expend time saving those who can be saved (Constraint: Return victims to duty as early as possible to support mission)	2.1.1 Conduct quick scan of victims and group/tag according to M (Minimal), E (Expectant), D (Delayed), I (Immediate) 2.1.2 Set up victim collection point 2.1.3 Group victim in sets of E, D, and I; return M to line 2.1.4 Assign "buddy" to monitor sets or individuals	Number of victims Basic information and pertinent history on each victim: name, blood type, religion Mechanism of injury Obvious injuries

FIG. 8.9. Task/subtask matrix expands the simple tabular task analysis.

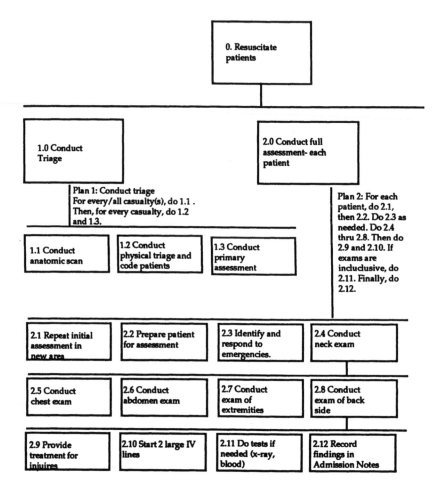

FIG. 8.10. Partial graphical task/subtask analysis of tasks a surgeon performs when multiple trauma victims arrive at a hospital (Format adapted from Kirwan & Ainsworth, 1992).

in a word processing, forms creation, or database package. It is similar in its structure to an object template. Figure 8.11 depicts a sample task frame.

This format can be very useful later, in the development and refinement of user interfaces that support the various tasks. For example, we place the task template for a task on the left-hand side of a posterboard or large piece of Kraft paper. On the right of the posterboard we place draft screens or storyboards that depict what the performer might look at and work with while completing the task. We place these posterboards around the room and conduct a focus group with performers. After we elicit feedback, we use it to refine the required functionality and screens.

Task: Conduct triage	
Performer:	Physicians
Description:	Sort patients according to the priority of their casualty/need
Goal:	Determine category for each patient and what should be done
Stimulus/cue:	Casualties/victims arrive
Location:	Trauma center or ER
Temporal aspects:	20–30 seconds per patient
Recipients:	Injured personnel Patient administrative officer
Resources:	• 2 physicians • 1 patient administrative officer • Nursing staff • Chief of ER in charge

Subtasks:	1.1 Conduct anatomic scan of patient
	1.2 Conduct physical triage
	1.3 Conduct primary assessment
	    1.3.1 Assess airway
	    1.3.2 Assess breathing
	    1.3.3 Assess circulation
	    1.3.4 Assess mental status

Equipment:	Vital sign sensors/monitors
Information:	Airway status Breathing patterns Respiratory rate Pulses Blood pressure Level of consciousness

FIG. 8.11. Sample task frame.

How to Complete a Hierarchical Task/Subtask Analysis

Regardless of the representation you select, to conduct a hierarchical task/subtask analysis you should start with a scenario on which to perform the analysis or observe and monitor task completion. Next, identify the goals and plans that drive the tasks, and the operations or tasks the performer completes to attain the goal. Continue decomposing the tasks into subtasks until you have enough information to meet your project goals. One of the nagging issues is "When do I stop decomposing tasks and subtasks?" Most researchers respond that decomposition should end when you are at the lowest level you need (McGraw & Riner, 1987; Kirwan &

Ainsworth, 1992). This level varies depending on your purpose. Is the purpose training, in which you will develop material to enable a performer to complete the tasks? Or is it system development, including user interface design? The level of specificity required for system interface design would be deeper than that required for training development.

The specific steps for the hierarchical task analysis process include:

1. Identify basic information: the goal, task title, task number, and description.

2. Identify other information you want to record, and build the factors into a matrix. For example, decide whether you need a basic representation or an extended one.

3. Encode task information into a table. For each task listed in the table, complete a decomposition that will enable the performer to detail the task more specifically. Determine the numbering scheme and the depth the analysis must represent. Break each task down into subtasks—separate units that contribute to the completion of the task.

4. Determine the apparent mental plan the performer seems to use to complete each task. Does the performer complete all subtasks for each occurrence of the task? Or is some decision making and selection required? (The graphical format shown in Fig. 8.10 depicts the plan of operation beneath the task.)

5. Stop when the level of detail is appropriate to your purpose.

6. Review the tabular format with the performers to ensure its validity, then complete a graphic representation if required. We find it useful to first build a table that links information in a textual format, as was shown previously in Fig. 8.8 or Fig. 8.9, and then develop the graphical version. This also enables us to double-check our analysis.

Operational Sequence Analysis/Diagrams

Operational sequence analysis represents a fairly specific level of the way a system functions. A *system* is the total work process, including human performers working within an organization, and the computer systems they use to complete tasks. For example, to identify all the components of a hospital-wide system to manage the efficient ordering and processing of lab tests, analysts would have to look at the total system currently in place, including the doctors, nurses, patient administrators, billing analysts, computer systems, lab orders, and so on.

Purpose

The purpose of an operational sequence analysis is to investigate the interrelationships in a system, between human actions and processing tasks (Enos & Tilburg, 1979). In the simplest terms, it enables us to identify the operations associated with a task and the order in which they are performed. Two primary variations are commonly used, which provide slightly different information—basic operational sequence diagrams and temporal operational sequence diagrams (Kirwan & Ainsworth, 1992).

How to Complete an Operational Sequence Analysis and Diagram

First, develop (by observation) or select an appropriate scenario to analyze. Decompose the scenario into work processes, then decompose the target work processes into top level tasks.

Next, select a representation that meets your project goals. Operational sequence analysis may be represented in a number of ways, which vary according to the level of specificity depicted, the intended use of the representation, and the ease with which they are constructed. Regardless of the differences, however, the purpose of all operational sequence representations is to indicate where and how interactions and sequences affect the performer's goal, and to do so within the context of the system's normal environment. The *basic operational sequence diagram* is a flowchart that categorizes operations into behavioral elements that are then represented by predetermined, standard symbols (Kurke, 1961). For example, the representations shown in Fig. 8.12 are used frequently.

If there are logical relationships between the inputs and outputs in this type of diagram, they are indicated by the way the lines connecting the elements are drawn. Lines between elements that are linked before entering or after leaving an element indicate an AND relationship. Separate

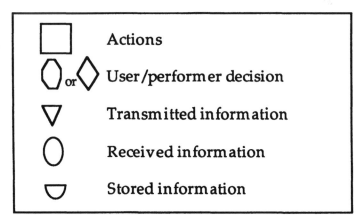

If operations are incomplete or contained an error,
the symbol is represented as half-filled.

FIG. 8.12. Standard symbols for basic operational sequence diagrams.

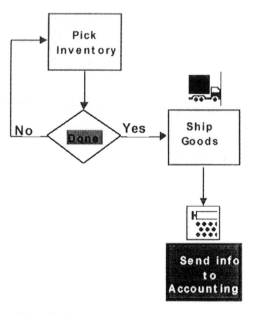

FIG. 8.13. Basic operational sequence diagram.

lines going into an element indicate an OR relationship. Figure 8.13 is an example of a basic operational sequence diagram.

The *temporal operational sequence diagram* is an effective way to structure tasks that are highly sequential in nature. As the name implies, this diagram focuses on the relation or effect of time on the completion of the task. It enables us to get at least an initial picture of the sequential nature of required performance and can help us isolate temporal performance requirements of tasks.

A temporal analysis may be continuous, from the start to the end of a series of tasks, or it may be compiled from several individual analyses, each of which represents a discrete task (Woodson, 1981). To construct a temporal analysis diagram, follow these steps:

1. Elicit information about the tasks and/or subtask names, as you would do for any basic task analysis. In addition, elicit information about accepted task sequence and the time required or allotted for task completion.

2. For each task listed, observe and record the *actual* time required for completion. Note start time, end times, and when the next task begins. Simply asking participants about task time requirements is ineffective because human performers are usually not accurate in estimating time completion, especially for compiled tasks. Note any discrepancies between required (i.e., according to procedures) and observed completion times.

3. Construct a chart or diagram to represent the timeline analysis. In many cases, a simple listing of the tasks with start and end times, total duration, and descriptions of events that constrain task completion can be shown in a table. If you want to represent time estimates, you should use a standard interval throughout the chart. The resulting chart could use graphic segments to represent both the span of time a task occupies and where within that task the next task begins.

For example, the diagram shown in Fig. 8.14 enables a visual check to identify extremely short tasks, tasks that comprise the largest chunks of time, and so on. As indicated in the figure, Task 1.6, "obtain/verify preprint information" is the most time consuming.

As depicted in Fig. 8.14 the major function or task number should appear as a label for the diagram, with subtasks and their associated descriptors and time estimations comprising the body of the diagram.

Cognitive Task Analysis

If the system to be developed is to eliminate performance problems, reduce training time, or enable the hiring of less qualified personnel, a cognitive task analysis may be required. So far, we have demonstrated techniques to illustrate primarily procedural, observable tasks. In many cases, this type of task analysis will be acceptable. However, for domains in which it is important to capture the cognitive dimension of tasks, standard task analysis may not be efficient. For example, a task that is represented as "Enter appropriate queue code" sounds simple enough. In observation, it appeared that this task was accomplished by selecting the appropriate screen

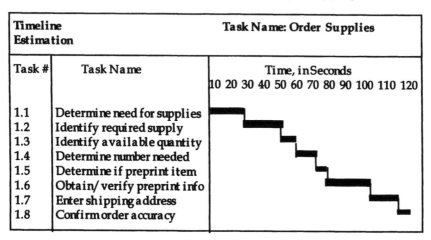

FIG. 8.14. Temporal operational sequence diagram, with standard intervals.

to review, asking the telephone customer some filtering questions, and entering the three-letter code in the appropriate field on the screen. However, further analysis revealed that this single task was the source of a very high number of errors, that it occupied a larger-than-expected chunk of the new hire training curriculum, and that it required the performer to memorize 30 codes, some of which were non-mnemonic and difficult to discriminate. Situations such as these lead analysts to consider using cognitive task analysis, which measures the impact of various cognitive factors on the completion of a task and attainment of a goal.

Purpose

A cognitive task analysis is likely to be completed for the development of intelligent interfaces or embedded performance support systems (McGraw, 1994a). Its primary purpose is to identify the tasks that can be categorized as:

- The most mentally challenging.
- Difficult to learn.
- Difficult or complex to do.
- Having high memorization requirements.
- Having the highest consequence of error.
- Presenting the most performance problems.

The end result is to determine which tasks require any of the following:

- A better user interface screen or enhanced interface functionality. For example, building a list of codes into the interface to reduce memory load during the task.
- An online, embedded "coach," advisor, or intelligent agent to assist in task completion. For example, a coach that helps an agent query a customer, provides fields for data entry, and searches to select the appropriate code for the performer, based on the context.
- An online, embedded "tutor." For example, if the performer views help files and uses a coach, but still does not fully comprehend the process, a tutor provides experience-based learning within the context of the job.
- Revisions to existing training curriculum.

How to Complete a Cognitive Task Analysis

To conduct a cognitive task analysis, first conduct work process observations and a basic task analysis. You will need an organized set of tasks, categorized by work process or job functions. Next, follow these steps (McGraw, 1994a), or modify them to meet your needs and purposes:

TABLE 8.8
Potential Dimensions for Cognitive Task Analysis Ratings

Select This Cognitive Dimension	To Measure
Complexity	Number of steps involved
	Number of parallel or simultaneous activities (mental or physical) involved
Difficult to Do	How hard it is to do the task
Difficult to Learn	How hard it is to learn the task
Consequence of Error	The impact of an error on the performer, corporate measures of success, or external entities (i.e., customers, supplies)
Memory Load Requirements	The memorization requirements of the task and their potential impact on task performance
Response Precision	Required exactness of the performer's responses
Workload	Number of decisions or outputs relative to the time allowed
	Length of time a level or response must be maintained

1. Determine the cognitive dimensions or factors against which you want to rate each task. Table 8.8 presents some of the dimensions from which you may choose. However, we caution you to keep the number of dimensions you select small, especially if you have many tasks to be rated.

2. Identify who will rate the tasks according to the dimensions you select. Will the raters be performers, or will the rating group include training specialists, supervisors, managers, and other support groups? For example, for a recent cognitive task analysis we selected a mix of performers, supervisors, and trainers and captured demographic data about them (e.g., years of experience, computer system use, special expertise, age, part-time vs. full-time, region of the country). We used this demographic data to ensure that the performer population we chose represented the total population of performers (using a stratified random sampling technique).

3. Refine the tasks that will comprise the body of the survey instrument. We use a Likert scale format for our cognitive task analysis (Likert, 1932), but other formats, such as the Thurstone scale (Thurstone & Chave, 1929), could also be used. The Likert format requires that each item be worded in a similar structural fashion, that it contain only one clause to clarify

what is being rated, and that respondents select from a set of responses to indicate how they would rate the item. First, we work with a small group of representative performers, supervisors, and trainers to fine-tune and refine each task item to be rated. It is extremely important that the terms, abbreviations, and acronyms used are well understood by all performers in the domain. Furthermore, this process enables us to eliminate redundant items or items that don't discriminate well (i.e., are not different enough from one another). For example, during our last cognitive task analysis we refined our list of tasks to be rated from an initial count of 120 down to 100 tasks. In addition, we clarified each remaining task.

4. Determine the number and content of the responses that performers will select to indicate how they feel about each task. We recommend that six levels of response be used, to minimize the probability that performers will develop a response set that makes them prone to selecting the midpoint. Try to select terms that fit together in a logical way, and indicate a progression along a scale. For example, we often use the following scheme:

| 0 = Not applicable | 1 = Not at all | 2 = Slightly |
| 3 = Moderately | 4 = Very | 5 = Extremely |

The higher the response, the more difficult, harder to learn, more memory intensive, etc. is the task. The lower the rating, the less performers perceive the task as being problematic.

5. Test a small set of items with the dimensions you want to rate and the rating scale you devised, to ensure that the tasks are easy to rate using the survey design you have selected.

6. Write the directions that will accompany the instrument. Give examples of how the tasks will be rated. For example, Fig. 8.15 provides some sample text we have used for directions:[4]

7. Compile the tasks, directions, and ratable dimensions into a usable format. Review the entire format and refine as needed. Table 8.9 illustrates a sample body of the cognitive task analysis instrument.

8. Plan and coordinate the session at which performers will complete the survey. We prefer to conduct this analysis in person to enable us to present the same directions to each group, to ensure that questions are answered consistently, and to enhance our response rate. If we must conduct multiple sessions at multiple sites to obtain the responses we need,

---

[4]This particular example is from a cognitive task analysis performed by K. McGraw while in the employ of RWD Technologies, Inc.

"This survey consists of 12 job functions and the key tasks you perform in completing each function. Read each task carefully, and think about it within the context of the job function. Place a number in each of the three categories to indicate how you feel about the task along that dimension. Use the scale that follows:

0 = Not applicable 1 = Not at all 2 = Slightly 3 = Moderately 4 = Very 5 = Extremely

For example, if the task takes up to a year to learn to do well, you might rate it as extremely difficult to learn. Place a '5' in the Difficult to Learn column to the right of the task. If it entails an average number of steps, average effort on your part, or is not particularly hard to do, you might place a '3' in the Complex or Difficult to Do column. If you have not had the opportunity to do the task, or if it does not apply to your job function, place a '0' in the column(s)."

FIG. 8.15. Sample directions for completing a cognitive task analysis.

we send each moderator/analyst into the field with the same presentation and materials, to enhance our reliability.

9. When the performers come to the session, present your goals and directions, give them the survey instrument, and facilitate the session. All completed instruments will then be brought back and entered into a statistics package.

10. Analyzing the results entails identifying the tasks that are rated consistently high. We look very closely at tasks rated "moderately" or above on any dimension. Statistics we use to analyze each task response include the mean (i.e., average) and mode (i.e., most often selected). For those tasks that are rated highly, we also may use analysis of variance to determine whether there are differences in the way that different groups of people (performers vs. managers, experts vs. novices, etc.) rated the task.

We use the results of our cognitive task analysis to make requirements and design decisions.

TABLE 8.9
Sample Cognitive Task Analysis Body Structure

#	Task Description	Difficult to Learn	Complex or Difficult to Do	Memory Intensive
1.1	Use appropriate phrase to request customer file.			
1.2	Locate appropriate file from Duplicate Listing screen.			
1.3	Add a new file if no file exists.			

ADVANTAGES AND DISADVANTAGES
OF WORK PROCESS AND TASK ANALYSIS

Work process analysis and task analysis are very useful tools for systems analysts and developers. However, disadvantages exist and must be managed.

Work process analysis and task analysis enable analysts to stay well grounded in the domain under investigation. They require that we conduct analyses within the context of the work environment, and in doing so, we gain a deeper understanding of the performance goals. Furthermore, work process analysis and task analysis are fairly reliable if guidelines for their use are followed. They reflect higher task and context fidelity than some other techniques because they so closely represent the real world. Work process analysis and task analysis fit well with a scenario-based development approach in which the focus is analyzing activities within normal and critical incidents. Finally, they enable us to represent their output in ways that are easy to understand and communicate with others.

Disadvantages do exist, however. First, if work process analysis is conducted solely from interviews, as opposed to interactive observations and scenario analysis, you risk basing your analysis on summary or descriptive information that may not be valid. Second, if you are not trained in task analysis, the learning curve can be steep. You must be able to determine the classification scheme that is most reasonable for the domain, and must manage complex concepts such as *task* that are not easily defined. You must also be able to select appropriate task verbs and develop task statements properly. Third, you must be able to observe, quickly identify what you are seeing, categorize the activity, and record it in a way that enables discussion with the performer. These nearly simultaneous tasks may require that two analysts work together during a session—one facilitating, observing, and questioning, and one recording observations. Fourth, you must be able to select appropriate decomposition representations, based on the goal and purpose of the analyses. These representations are time consuming to produce properly. Fifth, you must be able to work well with performers in their environment.

Finally, you must be able to decompose the tasks represented by the work process to the level of specificity required. A failure in any of these areas will render your analyses less than effective. Getting it right the first time is critical because the output of your work process and task analysis is used to conduct concept analysis and to begin identifying required components and other architectural elements.

Chapter 9

Eliciting and Analyzing
Domain Concepts

Object-oriented development, and particularly the Scenario-based Engineering Process, requires that development teams look beyond the simple surface level of functionality when developing requirements and specifications. When an analyst begins domain analysis for a complex system, he or she is often confronted by a seemingly unrelated morass of unfamiliar terms, objects, definitions, and applications. Concept analysis enables analysts to identify categories in the domain and discriminate among them using a feature set. Completion of this process results in the analyst's development of a "mental map" of the domain, which helps him or her understand the domain better. More importantly, conceptual products provide a visual abstraction of the domain that can represent the domain to systems developers and provide a bridge to specifications.

In this chapter we provide a rationale for concept analysis and define its role and use within the Scenario-based Engineering Process. Next we describe how to extract concepts from scenarios and sessions with performers, and through document analysis. Then we present techniques, such as concept maps and taxonomies, to organize and represent concepts. Finally, we address the task of refining the products of initial concept analysis activities.

This chapter is intended to provide a background in the use of conceptual elicitation or extraction and concept analysis, and their roles in the SEP domain analysis process. It enables analysts to meet the following goals:

- Recognize the criticality of adequate concept analysis during the domain analysis phase.

- Be able to extract concepts from scenarios and other primary elicitation sessions.
- Be able to extract concepts from domain documents.
- Analyze and represent concepts in visual formats that can be reviewed by performers and used by system developers.
- Facilitate the review and refinement of initial concept analysis products.

BACKGROUND

In the Scenario-based Engineering Process, domain analysis is critical during the creation of specifications. Domain analysis is the vehicle analysts use to increase their understanding of the domain by constraining its description. Analysts model the domain to help limit the domain description, and thereby the complexity, so they can draw conclusions by studying the models and not be overwhelmed by unconstrained data (Haddock & Harbison, 1994).

Most elicitation and domain analysis techniques, like work process and task analysis, enable analysts to model activities and entities within a domain. Others, like concept analysis, enable analysts to model the *organization* of a domain. Concept analysis provides a tool analysts can use to create representations that can be mapped more easily to object-oriented constructs, which in turn, can be analyzed and developed into the specification for the system (see chapter 2).

What Is Concept Analysis?

Concepts are symbols or abstract representations for common characteristics or relationships shared by objects, entities, or events that are otherwise different. They are building blocks of semantic knowledge, representing objects, attributes, events, relationships, and beliefs in a format that can be easily expressed and reviewed. Concept analysis enables the analyst to focus on the meaning, or semantics, *behind* data objects and entities. The semantics of a domain are best understood as the definition and relationship of one object in the domain to another. Concept analysis enables the analyst to investigate the domain in greater depth, defining in more detail the key concepts, their organization, constraints, contexts (i.e., tasks or processes in which they are used), and associated objects. By examining concept maps, taxonomies, or other output from concept analysis, analysts can easily convey the meaning of a concept or object through its position (Haddock, Kelly, Burnell, & Harbison, 1994).

Concept analysis has these primary uses:

- Enabling analysts and project team members to better understand and manage complex domains, which require more than a surface-level (i.e., declarative) understanding of critical objects and elements.

- Ensuring adequate coverage of a domain by providing a means to extract and represent in graphic formats the significant concepts used in scenarios, processes, and tasks. The resulting representations can be compared to multiple scenarios and reviewed by performers more easily than other formats to identify inadequacies, faulty associations, or incomplete definitions.

- Providing a bridge between other elicitation techniques and the development of domain languages and specifications. Creating the specification for the system involves constructing the domain description language and formulation of the architecture description language. Domain knowledge is analyzed through structured, conceptual, and scenario analysis techniques. Concept analysis identifies the significant concepts and provides a means to extract from these concepts requirements embedded in semantic knowledge.

Using Concept Elicitation and Analysis

Concept elicitation and analysis are secondary techniques. That is, analysts should not *begin* the elicitation of requirements for a complex system with these techniques. Instead, concept analysis is applied after initial domain analysis has been conducted using observations, scenario analysis, or task analysis.

Performers are usually uncomfortable with concept elicitation and analysis activities. They have difficulty thinking about the domain in terms of concepts, or may not even know what a concept is—both of which make the session awkward and possibly nonproductive. These types of sessions require performers to verbalize or otherwise express (by moving and grouping concepts) knowledge that is semantic in nature and thus, generally not available in short term memory. For example, to respond to the analyst's prompts during concept elicitation and identification the performer must (a) visualize the domain, (b) isolate primary concepts that have been highly abstracted, (c) decompile the abstraction, (d) recode the inherent meaning into appropriate terminology, and (e) convey that to the analyst. As they struggle mentally to retrieve conceptual knowledge and recode it into a format they can express, performers may experience feelings of inadequacy, uncertainty, and frustration.

Concept analysis sessions should be held *after* the output from other sessions has been prepared because it is much easier to extract and identify concepts from this material than to elicit them directly from a performer. If analysts stimulate concept elicitation and analysis using output from scenario, observation, task analysis, or process tracing sessions in which the performer participated, he or she is more productive and generally experiences fewer frustrations. A session report that details a scenario or task analysis provides a way to ground the performer to a working context.

Concept elicitation and analysis are less easily defined than other SEP techniques and may involve any of the following:

- Conducting sessions in which (a) concepts are elicited or extracted from scenarios, recorded observations, or tasks analyses, or (b) key domain documents are analyzed for semantic content.
- Conducting sessions in which concepts previously identified as important to the domain are defined and organized to depict relationships.
- Reviewing products from conceptual analysis to examine complexities, inaccuracies, or completeness, and refining the original visual abstractions or products as necessary.

Extracting Conceptual Information From Sessions

Concepts can be extracted in special concept analysis sessions, or by reviewing:

- Scenarios elicited earlier in the requirements gathering process.
- Observation session reports.
- Work process or task analyses.
- Process tracing session output.

As the analyst conducts the session or reviews output from previous sessions, he or she can work with a performer to identify concepts that should be understood in more depth to refine the requirements for the components and functions. The purpose of these sessions is to define, organize, and review or refine significant concepts in the domain.

CONDUCTING A CONCEPT ANALYSIS SESSION

Schedule, plan, and introduce the concept analysis session as you would any elicitation session (see chapter 3). It is particularly important that the analyst make the performer feel comfortable about this technique and the type of output expected. It is often useful to bring in sample concept maps, models, or other representations to help convey session goals.

During the session, analysts ask appropriate questions to elicit the type of information desired. Analysts should use a structured interview format (see chapter 7) to help focus the session and to avoid long-winded discussions of the domain that may not be pertinent. Questions should not only elicit identification of significant concepts, but also definitions for their attributes, relationships, and complexities. Questions such as those shown in Table 9.1 are useful in eliciting and analyzing concepts. Allow time toward the end of the session to review and refine the concepts, their definitions, and relationships identified during the session.

It is highly unlikely the analyst will be able to complete the identification and definition of all concepts' attributes, relationships, or complexities in any single session. In fact, for large, complex domains the analyst may succeed only in identifying the concepts in the first session and beginning the definition process, or may complete the identification and definition of concepts for one or two of the tasks in the domain. Additional sessions will be required to complete the definition of the relationship and complexities of concepts.

The sections that follow provide more information on accomplishing each of the session goals within the framework of concept elicitation and analysis sessions:

- Identifying or extracting concepts.
- Defining attributes.
- Defining relationships and organization.
- Defining complexities and refining initial definitions.

Identifying Concepts

Analysts use a structured interview format in the session to help focus the performer on the identification of concepts. Prior to meeting with the performer, analysts prepare a document that summarizes a scenario, observation, task analysis, or process tracing session.

Using this document, highlight the concepts that appear to be used in tasks and decision making. (We often use a highlighter pen to identify and mark them and sometimes accomplish this step prior to meeting with the performer.) If the domain is extremely complex, with multiple processes and functions, we suggest focusing on identifying and defining concepts used in one or two tasks during the first concept analysis session. Set up subsequent concept identification and analysis sessions focusing on other tasks.

As the performer identifies significant concepts, use a notecard, sticky note, or similar item to represent them and to help focus the performer on one entity at a time. Write the concept's name on the item. As the performer works with the concept, he or she manipulates this item. As the performer defines a concept, the analyst can place the corresponding item

TABLE 9.1

A Questioning Scheme to Elicit and Refine Concept Identification and Definitions

Identify	Define			Refine
	Attributes	Relationships	Complexities	
Let's brainstorm to identify the concepts used in this scenario.	How would you describe this item?	Circle or move into a group the items that somehow "go together."	When would items in this group be most important, versus items in the other group?	Are any elements or concepts missing from this group?
Name the concepts associated with this task.	What elements or attributes define this item or concept?	How would you use items in this group versus the items in the other group?	Why did you group them in this manner?	Can any of these concepts be combined and re-named to eliminate redundancies?
What objects and concepts are used in this process?	In what ways are the items alike?	In what scenario, process, or task would this concept (or group) be used?	Are there possible areas of overlap between these items or these groups?	Which additional elements or concepts might be added to the initial groups?
Label each group of concepts we have defined.	How or when is each item used?	Are there different types of these items?	Is there any item that can be a member of more than one of these groups?	Which elements do not seem appropriate for either the group or the label?
	Is there any relation between how the items are used?	Do any hierarchical relationships exist among these items?	In what other ways could items be grouped and on what factor, attribute, or rule are you grouping them?	As we review the scenario phases, identify any new concepts or groups we should add.
	Are any values associated with using this item?			

in a "complete" stack. The analyst asks questions to elicit adequate identification of the concepts and, if time allows, elicit their definition.

Defining Attributes

After at least some of the concepts have been identified, the focus changes to eliciting their definitions and discovering relationships among them. To begin the definition process the analyst presents previously identified concepts to the performer by one of the means shown in Fig. 9.1:

- Posting on the wall Kraft paper strips, on which are attached concept index cards and sticky notes on Kraft paper representing the domain or specific tasks or processes.
- Positioning index cards and sticky notes on the table in front of the performer, or on a white board in the room.
- Presenting the performer with a graphics program file that shows concepts as ovals, circles, etc., that can be moved around and annotated easily.

Next, the analyst uses questioning techniques and methods such as those that follow to extract a definition for each concept.

Simple Definition

Performers may find it easy to generate the definitions of very familiar, well-understood, and often-used concepts or concepts that represent significant persons, places, or concrete things. The analyst may need only to ask questions. The second column of Table 9.1 includes sample questions that help define attributes, such as:

- How would you describe this item?
- What elements or components make up this item or concept?
- In what ways are the items alike?
- How or when is each item used?

FIG. 9.1. Stimulating the definition of concept attributes.

- Is there any relation between how the items are used?
- Are any values associated with using this item?

Using Compare and Contrast

After the performer has completed simple definitions for each concept, ask him or her to compare and contrast specific concepts. The purpose of this activity is to isolate the salient features that define each concept. Salient features are the attributes that help discriminate between concepts and are key to forming more precise concept definitions.

Focusing on a selected pair or set of concepts the analyst asks the performer to compare and/or contrast them. The performer should be encouraged to state explicit reasons for the similarities and/or differences specified. Sample questions that analysts can use to stimulate comparisons and the definition of attributes and salient features include:

- How is Concept A different from Concept B?
- In what way are these Concepts alike?
- Is there any similarity in the way Concept A and Concept B are used?

Defining Relationships

Some concepts are very closely aligned with one another, whereas others are more unrelated. Once they have been exposed, conceptual associations or groups may be classified together under a broader umbrella that reflects all of the concepts.

Linking Concepts to Processes

One way to illustrate relationships among key concepts in a domain is to analyze scenarios (see chapter 5) and create a graphic that links concepts to different processes. Placement indicates the point in a process at which a concept comes into play. Such a representation enables analysts to distinguish between the different levels in a process and to understand and describe in some detail the relationships among levels for the purposes of process design, analysis, and control.

Conceptual graphs can be extended to represent generic physical processes. For example, Fig. 9.2 illustrates a process step within an emergency medical care process. In this step the emergency care giver must determine patient priority (e.g., Priority 1 requires immediate care while Priority 3 enables delayed care) to plan emergency interventions and to determine

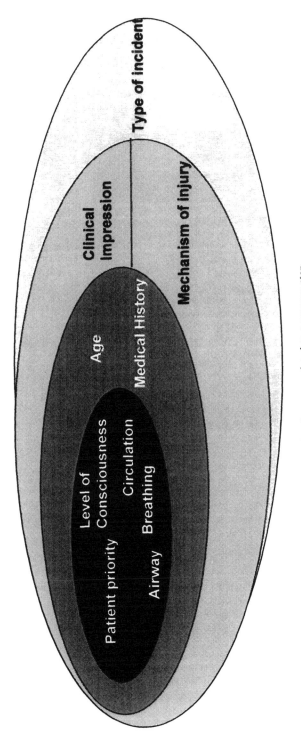

FIG. 9.2. Concepts related to a process step.

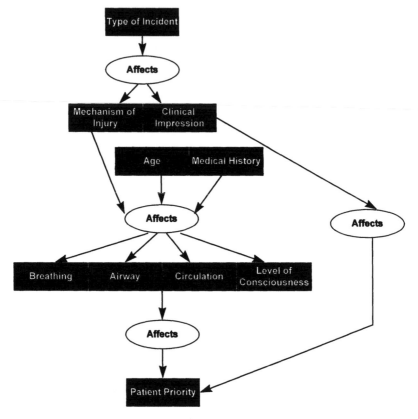

FIG. 9.3. Conceptual graph illustrating which concepts affect others during determination of appropriate patient care.

how (e.g., ambulance, helicopter) and where (e.g., community hospital, emergency room, trauma center) to transport patients.

Next, each primary concept represented in the process step can be graphed as a state description, which embodies a specific set of variables that affect one another. Figure 9.3 illustrates this extension.

As indicated, the type of incident affects the mechanism of injury and the initial clinical impression (i.e., electrocution by lightning, motor vehicle accident, etc.). The mechanism of injury and the patient's age and pertinent medical history all affect the status of the patient's airway, breathing, circulation, and level of consciousness. The values or readings determined for the patient's airway, breathing, circulation, and level of consciousness during the primary survey, combined with the provider's clinical impression, will impact plans for emergency prehospital interventions and plans for transport.

Creating Taxonomies

Conceptual clustering, a methodology for organizing and summarizing objects or concepts in a domain, is the process of grouping "exemplars" in logical ways (McGraw & Harbison-Briggs, 1989). According to Steep (1987), one way to cluster concepts is to present them in a hierarchy of categories or a tree structure. One way to describe conceptual identifications, clusters, and dependencies is the use of taxonomies, or basic classification systems. Shannon (1980) contended that the validity of any analysis depends on the use of a valid taxonomic model.

Taxonomies are fairly simple to develop when analysts work with performers using scenarios already gathered from the domain. For a hospital system designed recently, one step in the scenario required that the patient be sent to a medical department for treatment. However, there are multiple departments within large modern hospitals, each of which offers specialized treatment. Each department has a name, staff with varying levels of training and certification, and specialized equipment. Figure 9.4 illustrates a taxonomic structure representing types of departments (Hufnagel, Harbison, Doller, Silva, & Mettala, 1994).

To construct a taxonomy, the analyst must work with performers to identify, define, compare, and group elements, working within the following guidelines:

1. Identify the exemplars the analyst and performer believes are important to represent as conceptual clusters.

2. Define the classification process or heuristic that will determine how the concepts are clustered into the taxonomies. For example, "landscape plants" could be represented by a number of different taxonomies. One of these would use "types-of-plants" as the classification heuristic, resulting in groupings such as "flowers," "shrubs," and "trees," as shown in Fig. 9.5. This structure reflects group membership based on an "is-a" scheme. Other schemes include organizing groups based on shade versus sun, coloration, preferred exposure or climate zones, and so on. Other classification processes that could drive the taxonomy are *has-a* (organization based on attributes defining a concept) and *is-part-of* (organization based on component parts of entities).

3. Further define each of the significant concepts. This definition process includes associating a value with the set of attributes that define each concept (Galloway, 1987) and identifying the salient attributes, or those that define the group membership. The attributes and values will depend on the heuristic or classification process being applied. For example, sample attributes for "annuals," a member of the "flower" group are (a) they bloom, and (b) they last only a single season and must be replanted. The

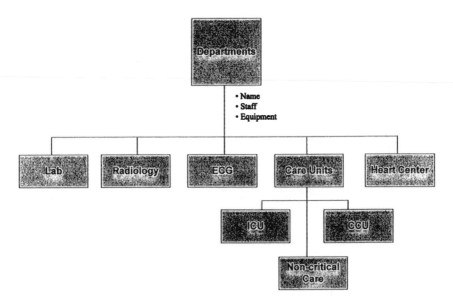

FIG. 9.4. Sample taxonomy for medical departments.

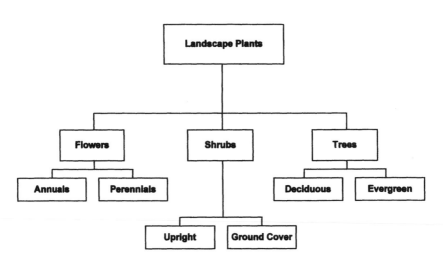

FIG. 9.5. Partial is-a taxonomy for landscape plants based on types of plants (adapted from McGraw & Harbison-Briggs, 1989).

second attribute is a salient feature that differentiates "annuals" from "perennials."

4. The analyst prompts the performer to refine initial groupings based on this definition process.

Using Repertory Grids

Repertory grid analysis is based on personal construct theory—the notion that each person functions as a "scientist" who organizes and classifies concepts and objects in his or her world (Kelly, 1955). Based on these classifications, the individual is able to construct theories of how the particular domain functions. Using this information, he or she is able to predict and act in the domain based on these personal theories. The classifications can be analyzed to identify associated constructs, which can then be presented using a repertory grid to represent the person's comprehension of a specific entity or concept (McGraw & Harbison-Briggs, 1989).

The repertory grid technique can also be used to represent a performer's organization of basic concepts in a specific domain. Although it is useful in complex domains involving decision making, the construction of these grids is less effective in simple procedural domains.

To develop a grid the analyst follows these basic steps:

1. Identify the concepts that need to be better understood or classified.
2. Work with an expert performer to elicit a set of constructs, or bipolar characteristics, that can be used to trigger the classification and organization of the selected concepts.
3. Stimulate the performer to provide a set of elements—examples that can be rated according to the constructs provided. For example, various chemicals would be elements that could be classified according to the construct "stable–unstable."
4. Ask the performer or a small group of performers to rate or rank each selected element according to the constructs that were provided. This rating can be facilitated by the use of a visual aid, such as a linear representation similar to those used in Thurstone attitudinal scales. Figure 9.6 illustrates a sample bipolar rating scale.
5. Present the ratings to a small group of performers or project task force to refine initial ratings.
6. Build a graph that visually represents the similarities and differences among the elements. Shaw (1981) and Hart (1986) suggested using factor analysis and cluster analysis; however, a spreadsheet application such as Excel® can produce useful results.

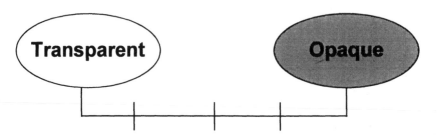

FIG. 9.6. Using a repertory grid methodology to rate and analyze degree of transparency along a bipolar scale.

Extracting Conceptual Information From Domain Documents

Analysts can extract major concepts, their definitions, and relationships from texts and training materials. Texts and training materials are a good place to start because they represent the author's attempt to present critical domain information in a logical manner. Additionally, the definitions and attributes associated with key concepts in this type of material are less subject to personal perspective and bias. Another location to review for concepts is the project glossary or dictionary that defines the terms and acronyms for the domain.

ANALYZING AND ORGANIZING CONCEPTS

After concepts have been extracted from sessions or domain documents, they can be analyzed and organized using any of the following techniques:

- Developing concept maps and models.
- Developing conceptual frameworks.
- Developing concept reports or "dictionaries."

Developing Concept Maps and Models

Concept models or maps are the result of a domain analysis in which the primary concepts in the domain have been identified and their relationships illustrated. They are relevant to a broad spectrum of problems in the representation and modeling of domains or knowledge within a domain.

Concept maps represent the ideas of a selected performer (or small group of performers) or set of domain documentation concerning the primary concepts and interrelationships in a domain. These maps enable the analyst quickly to become familiar with major concepts and relationships, reducing the amount of time required for document analysis.

Creating a concept map is similar to mind mapping, a technique often used in brainstorming ideas (see chapter 11).

Guidelines for creating concept maps are as follows:

1. Decide the level of specificity desired. For example, in a single session with a performer it will be impossible to "map" all relevant concepts.
2. Determine the formalism and notational mechanisms that will be used to map the domain.
3. Name the map. This name may become the central node in the map or its label.
4. Ask the performer to name the major entities involved. For example, in Fig. 9.7, the main name is Civilian Trauma Care and the performer thought of (and mapped) it according to Personnel, Activities, and Equipment.

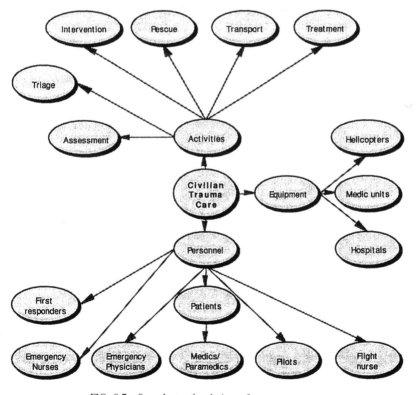

FIG. 9.7. Sample top-level view of a concept map.

5. For each entity identified, ask the performer to list the component parts, types, or other related concepts.

6. Guide the performer to remain at a certain level of specificity (i.e., you want to ensure you get all concepts identified at one level first in this session). If the performer starts to decompose a concept before you are ready, let him or her know you intend to explore each in more depth later, but want to focus on a particular level for this session.

7. Ask the performer to review the map and ask clarifying questions about areas that appear redundant or overlapping. Go around the map, focusing on each major entity, and ask the performer if there is anything missing or any changes he or she would like to make.

Another way to develop a concept map is to work from domain materials. For example, to create a concept map that illustrates clinical impressions that might be assigned to a recipient of emergency medical care, analysts could review print information that might include the sample shown in Fig. 9.8.

Using these definitions, analysts can organize concepts and develop a map that represents their interrelationships. Figure 9.9 is an example of

---

Firearm injury
    Includes firearm injury-accidental, assault, self-inflicted-intentional, and the use
    of handguns, shotguns, hunting rifles, and others.
Electrocution, non-lightning
    Includes accidents related to electric current from exposed wire, faulty
    appliance, high voltage cable, live rail, or open socket.
Electrocution, lightning
    Accidents related to lightning.
Inhalation injury, toxic gas
    Excludes smoke inhalation.
Smoke inhalation
    Smoke inhalation encountered in conflagration setting.
Respiratory distress
    Includes patients with respiratory distress who have spontaneous breathing and
    never suffered respiratory arrest.
Respiratory arrest
    Instances in which patient stops breathing.
Cardiac rhythm disturbance
    Includes rhythm disturbances noted on exam or monitor, when rhythm was the
    reason for care rendered.
Cardiac arrest
    All instances in which arrest occurred and death was pronounced or massage
    instituted.

FIG. 9.8. Sample entries from a domain document provide the foundation for a concept map.

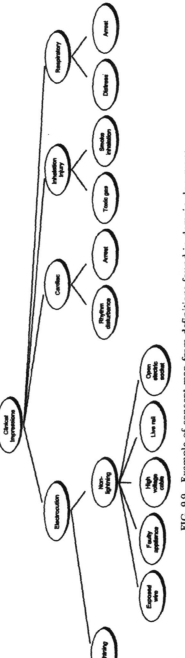

FIG. 9.9. Example of concept map from definitions found in domain document.

a map representing the relationship of concepts shown in the previous figure.

Developing Conceptual Frameworks

Conceptual frameworks are tools analysts can use to depict the elements that appear to be critical to understanding and applying concepts. McGraw & Harbison-Briggs (1989) adapted them from the frameworks Carney (1972) used to monitor and analyze communications.

Analysts can use a standardized framework template to analyze and represent categories of critical information used by performers working together in elicitation session to solve a problem, or presented in textbooks or training manuals. Sample categories of information that might be depicted in a framework include:

- Subject matter categories.
- Concepts or objects mentioned.
- Frequency of mention.
- Associated attributes and values.
- Length of discussion.
- Intensity of discussion (on a predetermined scale).

To use this tool to capture important concepts and their use, analysts should follow these guidelines:

1. Analysts meet with other team members and develop a standardized template or framework to use in representing concepts used in sessions or documentation. Later, the columns in the framework can become slot names in a concept dictionary. Figure 9.10 illustrates a sample structure for a conceptual framework.

Subject matter category	Concept	Attribute	Value	Frequency
Investment strategy				
	Investor	Age	• Below 30 • >31	7 in 15 minutes
	Investor	Family status	• Married & children • Married, no children • Single	6 in 15 minutes

FIG. 9.10. Sample structure for a conceptual framework.

2. Transcribe or review information from previous sessions to develop a conceptual framework that structures the analysis and isolates the subset of the domain under scrutiny. Assign different analysts different documents to analyze using the framework. The framework provides a mechanism to compare and synthesize information gathered from diverse sources.

3. Present the resulting conceptual frameworks to a small group (i.e., 2–3) of performers for review and refinement.

4. Use the resulting framework as input to domain dictionaries, domain models, and so on.

Developing a Concept Report

The vocabulary performers use represents major concepts within the domain. McGraw and Harbison-Briggs (1989) suggested that engineers and analysts use a project-wide tool such as a visual concept dictionary or report (Carney, 1972) in addition to a glossary of terms. The concept report provides a mechanism to explicate concept definitions, types, components, and uses. In addition, it can include concept diagrams, such as maps, that provide visual abstractions of the primary concepts in the domain. This report is based on the notion that concepts may be grouped by common elements of reference according to some logical anchor. Because we work from a task analysis to identify concepts used in the completion of tasks, our concept report is usually organized by tasks. Figure 9.11 presents an example of one entry from a concept report.

Guidelines to produce a concept report include:

1. Conduct and document an observation, task analysis, or process tracing session. Then, using the output from the session, highlight the concepts that appear to be used in tasks and decision making.

2. Using a frame-like structure, create an entry for each concept identified. Typical entry "slots" that could be documented include: concept name, task used in, definition, used by, types of, components of, and so on.

3. If a concept has "children" component parts or types, a visual representation might be helpful and can be created easily with tools such as Inspiration.

4. As concepts for each task or process are organized, eliminate redundancies by referencing previously created entries.

5. Review the project glossary to ensure consistent definitions for concepts and to add new terms to reflect concepts identified during the session.

6. Present the entries to a performer for verification and refinement.

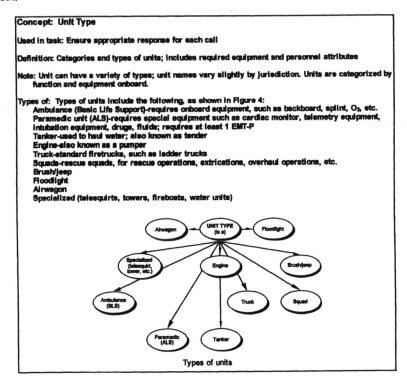

Concept: Unit Type

Used in task: Ensure appropriate response for each call

Definition: Categories and types of units; includes required equipment and personnel attributes

Note: Unit can have a variety of types; unit names vary slightly by jurisdiction. Units are categorized by function and equipment onboard.

Types of: Types of units include the following, as shown in Figure 4:
    Ambulance (Basic Life Support)-requires onboard equipment, such as backboard, splint, O₂, etc.
    Paramedic unit (ALS)-requires special equipment such as cardiac monitor, telemetry equipment, intubation equipment, drugs, fluids; requires at least 1 EMT-P
    Tanker-used to haul water; also known as tender
    Engine-also known as a pumper
    Truck-standard firetrucks, such as ladder trucks
    Squads-rescue squads, for rescue operations, extrications, overhaul operations, etc.
    Brush/jeep
    Floodlight
    Airwagon
    Specialized (telesquirts, towers, fireboats, water units)

Types of units

FIG. 9.11. Sample entry from a concept report organized by tasks from scenarios.

Domain Dictionary

One of the possible outputs of concept analysis in SEP domain analysis is a domain dictionary whose entries extend domain concepts, classes, instances, attributes, or relations, and whose entries contribute to the domain model. Each domain dictionary entry transfers origination and other historical information with the semantic definition of the entity. Typically, a domain dictionary entry depicts the following:

- Name of the entry.
- Name of the context from which it was extracted.
- Type of entry—concept, class, instance, attribute, relation.
- Session or document from which it was derived.
- Informal description or definition.
- Specific known dependencies.
- Example usage.
- Name of the analyst creating the entry.

- Date entered.
- Name of the person modifying the entry.
- Date modified.

The most efficient format in which to organize a domain dictionary is as a database. The database enables all analysts to input entries into a single location. Analysts check for the existence of an entry prior to inputting it, thereby reducing redundancies. If an entry is found, the analyst reviews the entry's information and works with the originating analyst to modify it to fit the new context, if needed. If not, the analyst inputs the new entry. Figure 9.12 presents an example of a domain dictionary database designed in FileMaker Pro.

Information about entries can be printed in a variety of formats, including:

- Entries sorted by type.
- Entries and their definitions.
- Selected entries and their definitions.
- Entries with creation and modification dates.
- Entries with source derived from.
- Entries by analyst.

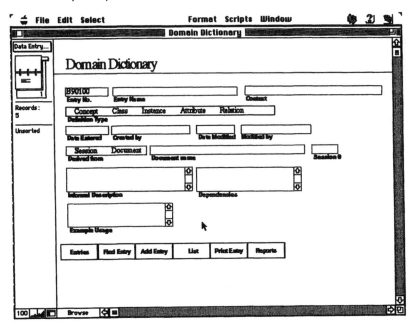

FIG. 9.12. Sample domain dictionary form.

Analysts can easily update and maintain the dictionary in this format. Reports may be obtained on a scheduled basis and organized by context, analyst, type, and so on. Reports such as these enable easy review and refinement, as well as providing an audit trail from each entry to its source.

DEFINING COMPLEXITIES AND REFINING CONCEPT IDENTIFICATION AND DEFINITION

Rarely are initial concept definitions and relationships final. After they have been identified, analysts must work with performers to define complexities that impact their accuracy and completeness, and to refine concept identification and definition.

Defining Complexities

Defining complexities requires that concepts be previously identified and at least partially defined. The existing definitions and visual abstractions are used to stimulate a performer to retrieve more in-depth information about the concepts or their groupings. As an analyst works with a performer to extend the current definition of concepts, their attributes, and organization, he or she should address the following:

1. Which concepts are shared across multiple contexts, tasks, or processes? After they have been identified, investigate their definitions and attributes to ensure they are appropriate for each context. For example, while working in the emergency medical care domain, we discovered that the definition of patient prioritization varies between the military and civilian domains.

2. When are the concepts in one group important versus the concepts in another group?

3. Concept groupings or relationships reflect a specific heuristic or process. In what other ways can concepts be organized?

Refining Concept Identification and Definition

Concept maps, models, taxonomies, and definitions represent a performer's perception and mental representation, and thus are subject to inaccuracy based on individual perception and bias. Other errors may occur as the analyst translates conceptual, semantic information into a visual abstraction. Consequently, the output of conceptual analysis and representation must be reviewed and refined. This process includes reviewing conceptual analysis products with additional performers to refine definitions and relationships, and examining them from the perspective of new scenarios (representing both standard process and complex process).

First, review conceptual analysis products with multiple performers either individually, or in small groups. These reviews must include a presentation of the scenario or context from which they were extracted, defined, and organized. The goals of refinement activities are to determine if:

1. Any elements, concepts, or groups are missing.
2. Any concepts can be combined and renamed to eliminate redundancies.
3. Additional elements, concepts, or attributes should be added.
4. Any concepts do not seem appropriate for the group or label applied to the group.

We recommend more than a discussion of the products to ensure the participation of the performers. One useful technique to elicit comments is to post the concept maps, models, taxonomies, and definitions along the wall of a conference room. Provide performers with sticky notes, and encourage them to attach comments to the products to provide additional definitions, constraints, or corrections. After the comments have been posted, the analyst facilitates a discussion of the critiques and gains consensus and closure (see chapter 11).

After the conceptual analysis products have been refined, convene a small group (e.g., 2–3) of performers. Conduct a test "walkthrough" of the maps, models, and definitions against at least one new scenario representing a "standard" process and one representing a more complex process. Concepts originally thought to be unimportant may become more important in specific contexts. In addition, concepts organized in one fashion may reveal a different relationship in another context.

The refinement of our understanding of a domain continues throughout the development of requirements and specifications for a system. Defining domain concepts and their interrelationships provides a set of data with which performers continue to interact to refine and clarify the initial representations.

Chapter 10

Using Process Tracing to Analyze the Problem-Solving Process

The task of analyzing decision making or problem solving is vital if systems are to support these business-critical skills. However, to meet this goal the analyst must be able to identify, elicit, and analyze requirements embedded in episodic and semantic knowledge.

In this chapter we present decision process tracing and protocol analysis, paired techniques that enable analysts to trace and study a performer's decision-making processes. These techniques provide a framework within which the analyst can focus on higher level (i.e., semantic) thinking. After studying a decision maker's responses to tasks and problems with which he or she is presented, the analyst can compile heuristics, alternatives, attributes, and attractiveness values for the decisions made and problems solved.

The effectiveness of decision process tracing can be enhanced by the analyst's preparedness, session management, analytical skills, and follow-up (McGraw & Harbison-Briggs, 1989; Waldron, 1985). First, analysts should complete domain familiarization activities and concept analysis sessions in the related area. This provides them with a background that enhances their ability to recognize decisions and their relationship to the concept's attributes. Second, analysts must be trained in setting up the problem, monitoring its solution, and analyzing and compiling the results. Third, decision process tracing produces a protocol (written or video) that analysts must carefully construct and analyze. Fourth, they must conduct a follow-up protocol analysis session (in the form of an interview or discussion) to refine and clarify the analysis of decision making information.

In presenting information on techniques to trace and analyze a performer's decision-making and problem-solving processes, we first address the notion of expertise by examining its components. Next, we define problem-solving and decision-making knowledge. Third, we describe the effective use of decision process tracing techniques, and outline the activities involved in protocol analysis. Finally, we discuss special considerations that must be addressed when using these techniques.

We suggest the use of decision process tracing and protocol analysis as techniques for acquiring decision-making or problem-solving knowledge. The chapter is designed to help the reader meet the following goals:

- Recognize the usefulness of verbal reports of decision-making and problem-solving information.
- Understand the link between the use of scenarios and these techniques.
- Select appropriate decision process tracing methodologies and techniques.
- Manage process tracing sessions, carefully monitoring the technique and the decision maker's response to it.
- Determine an appropriate analysis framework for studying and compiling this type of information.

WHAT IS EXPERTISE?

What is *expertise*? What is an *expert*? As common as these terms are, they are difficult to define. Expertise could be defined from a skills viewpoint by persons considered experts in a domain. This assumes that if we identify experts in a particular domain and examine their behavior, we will be able to define expertise for that domain. A similar, commonly accepted definition is that an expert is a "specialist" in a particular domain (Cooke, 1992). Typically, researchers have used experience as a measure of expertise (Shiffrin & Schneider, 1977). However, expert performance does not *always* correlate with experience.

One way to define expertise is to characterize its components. Figure 10.1 details some of the commonly mentioned components of expertise.

Many researchers believe that expertise is born out of having the right kinds of experience working in a domain or field. They conclude that performers gradually grow into experts. As performers begin working in a domain, their knowledge about the domain is encoded, or translated and stored as *declarative* knowledge (e.g., names of devices, basic facts about devices, etc.). They begin to learn general problem-solving procedures that enable them to operate on and apply their declarative knowledge about the domain. As they apply these procedures they refine them from general problem-solving procedures to include specific rules about devices and components, and specific problem-solving procedures for certain situations.

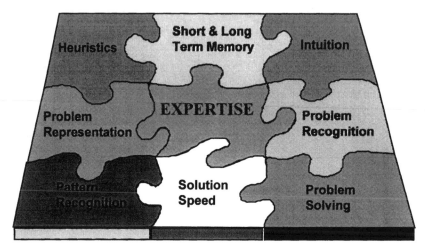

FIG. 10.1. Components of expertise.

After they have solved numerous problems, new ones can be analyzed and compared to previous situations. This is accomplished using structural analogies (Anderson, Farrell, & Sauers, 1984), which enable them to compare features and structures of new problems with old ones to aid in selecting (or modifying) an appropriate problem-solving approach. As performers experience more problems, they are believed to compile more complex, complete knowledge structures containing not only declarative knowledge, but also semantic, procedural, and episodic knowledge. (See chapter 4 for types of knowledge.)

Expert Versus Novice Problem Solving

Much of our understanding of expertise is communicated by comparing expert to novice behavior. Table 10.1 presents the characteristics of experts that are generalizable across a variety of domains (Glaser & Chi, 1988).

Experts differ from novices in that they have not only a solid foundation in declarative and procedural knowledge of their domain, but also have structured and compiled it so that it is easy to tap and apply. Not only do they know more, therefore, but they also have refined and organized their knowledge better. Primary differentiators between expert and novice performers include the following (LaFrance, 1989; McGraw & Westphal, 1990), each of which is discussed briefly in the sections that follow:

- Perception of problems and solutions.
- Goal-oriented thinking.
- Organization of knowledge.
- Episodic knowledge differences.

TABLE 10.1
Characteristics of Experts Generalizable Across Domains

1. Experts excel mainly in their own domain.
2. Experts perceive large meaningful patterns in their domain.
3. Experts are faster than novices at performing the skills of their domain and solve problems quickly.
4. Experts have superior short-term and long-term memory.
5. Experts see and represent a problem at a deeper level than novices.
6. Experts spend a great deal of time analyzing a problem qualitatively.
7. Experts have strong self-monitoring skills.

Perception of Problems and Solutions

Researchers believe experts and novices perceive problems and their solutions differently. A key difference is that experts examine not only the present problem, but *spend time initially analyzing and categorizing the problem* according to a particular type. Then, depending on the categorization, they select appropriate solutions. Novices, however, are more likely to begin to respond immediately to the new problem based on its *surface features*. In some instances, novices "don't see the forest" (i.e., the problem type and appropriate solutions) "for the trees" (i.e., the surface structure of the problem). Figure 10.2 illustrates this difference.

Goal-Oriented Thinking

Experts approach problems with a mental schema for their solution and exercise goal-oriented thinking in solving them. When presented with a problem, for example, they are far more likely than novices to view the problem-solving activity in terms of a plan they will put into effect. Novices are more focused on the consideration of single events. If asked to analyze their problem-solving behavior after an incident, they will often report each step as it happened, in a sequence, rather than thinking of the solution as a higher level plan comprised of a set of activities.

Organization of Knowledge

Experts have *more functional knowledge* than novices (LaFrance, 1989; McGraw & Westphal, 1990). However, experts don't just have better memories than novices; their knowledge is organized better to facilitate solutions. The contents of an expert's knowledge surpasses that of novices; however, the important difference seems to be that an expert's knowledge enables him or her to activate a higher level schema that may be relevant to a solution. This schema seems to contain more procedural knowledge and

FIG. 10.2. Novices often focus on surface features, rather than first categorizing the problem and selecting the solution.

knowledge about the conditions for applying the main principles underlying a given problem (Chi, Glaser, & Rees, 1982).

Another organizational difference is that expert knowledge appears to be "chunked" differently than novices' knowledge. Individuals could not use their knowledge efficiently and effectively if it were stored in individual pieces. True expertise is chunked, or organized, in knowledge structures. An expert's chunks also appear to contain *more complex knowledge* that is better integrated than that held by a novice. That is, an expert has more extensive and complex knowledge than a novice.

Episodic Knowledge Differences

Experts have *more extensive episodic knowledge* than novices. As performers experience events in the domain, they begin to turn previously unrelated facts into expert knowledge (Kolodner, 1983) that is coded into episodic knowledge. Episodic knowledge includes (a) the memory of situations experienced and organized according to characteristics of the event, problem types, and solutions, and (b) memories of how semantic, procedural, and declarative knowledge was selected and applied.

Novice decision makers evolve into expert decision makers as they encode episodic knowledge through experiences (LaFrance, 1989). Each additional episode has the potential of providing exemplar cases from which to reason in a new situation, as Fig. 10.3 depicts. As the new case

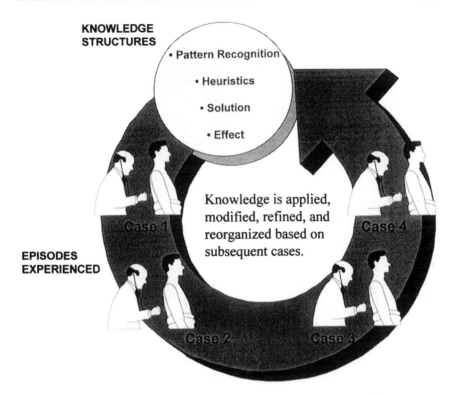

KNOWLEDGE STRUCTURES

- Pattern Recognition
- Heuristics
- Solution
- Effect

Knowledge is applied, modified, refined, and reorganized based on subsequent cases.

EPISODES EXPERIENCED

Case 1 Case 4 Case 2 Case 3

FIG. 10.3. New cases provide the opportunity to refine knowledge and add more exemplar cases to use in analyzing new situations.

is analyzed and encoded in the decision maker's memory, existing knowledge structures are elaborated and modified (Kolodner & Kolodner, 1987). This process continually refines the decision maker's knowledge structures, and helps the expert respond to novel situations.

Impact of Expertise on Requirements Elicitation

The complexity of the expertise that decision makers apply to solve a problem or make a decision impacts the analyst's ability to elicit, document, and detail related requirements. The following guidelines should be addressed to ensure that the requirements elicitation process for decision making and problem solving yields expertise by tapping both semantic knowledge (e.g., heuristics, mental models, etc.), and episodic knowledge.

1. *Analysts must elicit problem schemas and goals, not just problem-solving and decision-making steps.* It's not enough to be able to identify the steps a decision maker takes. True expertise begins with how the decision maker

views and recognizes the problem to enable him or her to apply the appropriate problem-solving schema and thereby, use the most efficient problem-solving approach. This schema, which is a mental representation of the problem-solving approach or plan, includes goals that must be met for the problem to be solved. Being able to identify these goals is critical because they stimulate the actual tasks and steps the decision maker takes to solve the problem.

2. *Analysts should immerse the decision maker in a problem-solving activity to focus on how expertise is applied and used, rather than simply eliciting facts and rules.* An interview will yield declarative information about facts and rules the decision maker uses. However, this information may be incomplete or inaccurate. Decision makers may neglect to mention pertinent information because it is too deeply embedded, or they may assume the analyst already knows it. As the decision maker talks about a specific area, he or she may forget to mention another step in the process, or area of consideration. Furthermore, interview sessions do not enable the analyst to monitor a "test" of their application. By presenting the decision maker with a simulated problem to solve, the analyst has more control of the problem-solving context and immerses the decision maker in the activity. The decision maker can focus on the problem and, within the solution process, can point out facts and heuristics used, as well as explain procedural or cognitive steps that were taken. Expert decision makers also are usually more comfortable in this more familiar environment. The information yielded is more accurate and complete (to the extent that the problem selected represents reality), and alternatives considered but not chosen can be discussed within the framework of attributes of the actual problem.

3. *Analysts should seek out chunks of expertise and attempt to understand how the expert decision maker has organized them for use.* If chunking is one of the differentiators between novice and expert decision makers, analysts must be interested in how specific chunks of expertise appear to be organized. Understanding this organization can yield invaluable clues to help in defining user interface design, knowledge-based support, and in identifying how we might better support novices toward their growth in expertise. As all analysts know, there are many ways to organize conceptual schemas during concept analysis activities (see chapter 9). Doing so in a way that is congruent with how expert decision makers think about the problem and the means to solve it enables us to design a more powerful, performer-centric system.

4. *Analysts should stimulate episodic knowledge.* Episodic knowledge is a natural outgrowth of having solved a number of different problems using many approaches (albeit with varying degrees of success) over a period of time. To each new episode, the decision maker brings the facts, heuristics,

conceptual organization, and experiences learned from previous episodes. Using these episodes, we identify similarities and differences between attributes of the current problem and those of previous problems to select reasonable approaches and solutions. Previously solved episodes contain expertise. Therefore, analysts should elicit them in the form of scenarios and ask the decision maker to identify the important attributes, goals, constraints, solutions, and a rating of the effectiveness of the solution during sessions (see chapter 5).

CONDUCTING DECISION PROCESS TRACING SESSIONS

Decision process tracing is any of a set of techniques that allows analysts to determine a decision maker's train of thought while he or she completes a task or reaches a conclusion (McGraw & Harbison-Briggs, 1989). It is an effective suite of techniques when analysts need to investigate information a decision maker uses to solve problems, and how he or she processes that information.

Protocol analysis is the process of analyzing the protocols, or verbal reports generated by a decision maker during a process tracing session. It has been used in numerous fields (e.g., psychiatry, anthropology, training) to generate and study data reported verbally. Newell and Simon (1972) reported using protocol analysis extensively to study people's general problem-solving approaches and the specific operations they used to move from one knowledge state to another.

Figure 10.4 illustrates decision process tracing and protocol analysis. The sections that follow detail the phases or activities involved in this pair of techniques.

Selecting the Problem and the Decision Maker

To ensure that the information from a session is most useful, the analyst must first select a good problem to solve or decision to make, identify the support material needed, and select an appropriate decision maker.

Decision process tracing sessions actively involve a decision maker in a task that includes solving problems, troubleshooting, and making decisions. Consequently, one of the most important presession activities is the selection of appropriate problems or cases to solve. This often requires working with an expert to define the type of problems or cases desired and any particular features, attributes, or constraints that should be used as criteria to select problems. The expert searches through cases and archival data

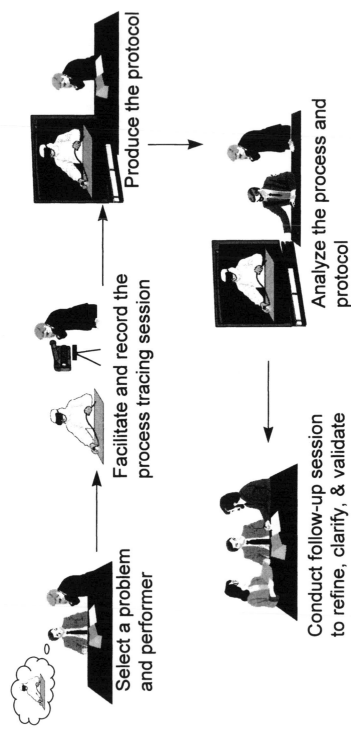

Select a problem and performer

Facilitate and record the process tracing session

Produce the protocol

Conduct follow-up session to refine, clarify, & validate

Analyze the process and protocol

FIG. 10.4. Eliciting requirements related to decision making usually requires the use of at least a pair of techniques—one to reveal the process and one to analyze and validate the output.

to select problems that meet the criteria. Next, the analyst and expert may need to prepare support materials (actual devices in various states, test data, patient records, etc.) the decision maker will use to solve the problem. Additionally, the problem or case may require some modification to enable it to be solved in the time available, using the selected technique. Analysts and experts must plan ahead—problem setting activities can be quite time consuming, especially if multiple levels of approval are needed to access and/or get permission to use actual cases and case data.

Based on the type of problem selected, analyst and customer liaisons or experts help identify an appropriate decision maker with whom the session will be conducted. Characteristics that are desirable for these sessions include:

- Extensive expertise in area(s) of the domain critical to the selected problem.
- Procedural experience and experience serving as a mentor or trainer of apprentices.
- Self-confidence in ability to solve similar domain problems.
- Ability to be self-analytical, not based on personal biases or rationalization, but on accepted protocol and heuristics for solving similar problems.

Conducting the Session

During the interview the analyst is active, asking questions, transitioning to new topics, and probing. In contrast, in the process tracing session the analyst's main activities are problem setup, observation, and recording the performer's actions. Here the analyst sets the stage, presents goals and expectations, gets the decision maker started, and then observes. He or she records the actions and verbalizations for later review, analysis, and follow-up interviews. Figure 10.5 illustrates the flow of information and communication in this type of session.

"First, I'll analyze the spreadsheet.
Then I'll look for exceptions to what I expect.
Here is one, in the beer and wine sales. It is too low.
Now, let's check the numbers."

DECISION MAKER   **ANALYST**

FIG. 10.5. Flow of information and communication in the decision process tracing session.

The analyst follows the suggested guidelines for planning any elicitation session (see chapter 3). In addition, however, recognize that although decision makers may know what to expect from an interview, they will not know what a "decision process tracing" session is. It has been our experience that few decision makers have ever experienced this type of session before. Consequently, the selected decision maker may feel some trepidation about the session. He or she may feel threatened, fearing that some inadequacy or fallacy in his or her thought process may be exposed. For example, techniques that constrain the amount of information presented require the decision maker to "solve" the problem with less data than might normally be available. This may result in discomfort as he or she struggles with incomplete information. It is the analyst's job to thoroughly explain the scenario or problem and its goals.

Using Techniques to Reveal Problem Solving

At least four techniques are commonly used to stimulate the revelation of decision processes during a process tracing session. Table 10.2 (McGraw & Harbison-Briggs, 1989) compares various techniques for eliciting and observing problem-solving activity, summarizing the strengths and weaknesses of each. Each technique is investigated in detail in the sections that follow.

Constrained Information Sessions

Constrained sessions are those in which the scenarios or tests that are used have been modified so that one or both of the following has been limited: (a) the amount of information provided, or (b) critical features such as time. Solving a problem without all of the pertinent data stimulates the decision maker to think strategically and to select the heuristics and information that are most useful. Constrained sessions require the decision maker to work with a limited set of information under a strict time limit. These methods are effective in revealing the problem's most salient features and the strategies the decision maker uses to seek a solution. The analyst can more easily isolate specific priorities and primary alternatives that are considered, along with the attributes that made them desirable in the specific situation. The goal is to enable the analyst to focus on the *strategies* the decision maker uses, as opposed to merely the factual information used, and to identify what the decision maker considers critical data for the solution.

To conduct a constrained information process tracing session, the analyst must first work with an expert to fashion a problem scenario or task that can be solved within the session's time limit. Next, an expert should assist the analyst in selecting pivotal information that can be removed from the problem or scenario to stress the solution and reveal the expert's strategy. For example, if the expert will be required to diagnose a skin

TABLE 10.2
Comparison of Problem Solving Observation Techniques

Technique	Strength	Weakness
Constrained Information	• Reveals salient features of a problem	• Discomfort on the part of decision makers
Constrained Solutions	• Reveals strategies	• Requires careful task selection • Requires restriction of appropriate variables
Simulated Scenarios	• Use of archived data • Not "real time" • Can interrupt solution process	• Requires extensive access to archival data • Requires assistance of expert to formulate scenario
Episodic Analogies	• Stimulates use of past experiences • Provides insight into salient features of a problem	• Requires decision makers to discuss past problem solving • May interfere with other strategies
Difficult Case Analysis	• Access to data that is infrequently used • Access to methods of combining and prioritizing data	• Difficult to plan • Requires decision maker involvement to select or record cases

disease, the problem may be constrained by *not* providing critical diagnostic information such as photos of the rash or biopsy reports. When the task is presented to the expert, the analyst explains that some information intentionally has been removed and that the goal of the session is to identify decision strategies and approach.

In one constrained information session, the analyst and an expert decision maker selected a problem in the domain of dermatology and a case in which a particular case of skin disease, dermatitis herpetiformis, was diagnosed. They reviewed the existing archival data (e.g., photos of the rash, description of symptoms, onset, etc.) and decided not to provide biopsy reports. Another decision maker was then asked to use the data provided to estimate an initial diagnosis. The analyst was able to isolate how the decision maker reviewed the data, grouped information, and approached the problem. When the decision maker requested a biopsy, he specified what he suspected and the additional information he would need to receive in order to confirm or reject his initial hypothesis.

Constrained Solution Sessions

Constrained solutions require the analyst to remove or manipulate a pivotal variable (e.g., time) in an otherwise familiar problem. To conduct a session with this method, the analyst presents the problem to the expert and tells him or her that a particular variable has been compressed for the purpose of isolating the factors that seem *most* important in determining a solution. The analyst specifies a window of time in which the expert can analyze the data. For example, he or she may allow the expert only 5 minutes to review data provided and formulate a plan. When that time is up the expert is expected to describe the plan and reasoning behind its selection. In follow-up interviews the analyst can clarify the most critical decision factors and explore the effect of imposed constraints on the decision.

In one case, the analyst and a decision maker selected a problem in the domain of dermatology to explore how experts diagnosed skin diseases. They chose a set of archived data detailing the diagnosis of dermatitis herpetiformis. They presented the problem to a decision maker, including photos of the rash, patient description of the symptoms, information about onset and prior treatment, biopsy, reports, and patient demographic data. They asked the decision maker to examine the information provided within 3 minutes. After that time the materials were removed and the decision maker was allowed to spend 5 minutes formulating hypotheses and supporting information for discussion. In follow-up interviews the analyst probed the decision maker's reasoning, critical decision-making factors, and the effects of the imposed constraints on the decision.

Simulated Problems

Scenarios or problems can be simulated using actual data from a task or problem that has already been solved. Hoffman (1987) described a version of this technique in which a common task is performed using archival (i.e., real) data. This technique has been used with physicians, for example, to examine how they would diagnose a particular class of illness. Given an "illness" or set of symptoms, physicians are asked to use real data to "guestimate" a diagnosis. The value of simulating the task, rather than just observing a real episode, is that analysts are able to "stop the clock" during the session and probe for reasoning strategies or information priorities.

To use this method, the analyst selects a familiar task and extracts the actual data that comprised the problem. This data is presented to a decision maker and described as real data from a similar case. The decision maker works through the data toward a solution and may be asked to talk about what he or she is doing. The analyst may probe the decision maker about a specific point, strategy, data being used, or current priorities. For example, during the development of requirements for a system that would do route planning, the analyst gathered a pilot, a navigator, and an intelligence specialist together. These decision makers were presented with archival data, including terrain and weather information, mission goals and requirements, known enemy activity in the area, and an initial mission plan. They were then given updated information and asked to replan a route through the territory.

As they worked together the analyst listened to and recorded discussions, noting questions asked and information that appeared to be pivotal. At the end of the session the analyst probed to identify specific information the experts were using to make decisions and determine the new route.

Upon completion of the session the analyst may wish to compare the original solution to that obtained by the current decision maker, or to investigate solution differences with the decision maker or a panel of decision makers. Although neither the original solution nor the one derived in the session may be all right or all wrong, a comparison of features, parameters, decision points, and rules applied may enable the analyst to view the problem from different angles and identify requirements based on this additional information.

Episodic Analogies

This method draws on decision makers' tendencies to use analogies to other problems or situations when confronted with a new problem. Researchers (Eberts & Simon, 1984; Hoffman, 1987; Klein, 1987) have contended that when called on to solve a problem, experts draw analogies to cases or situations with which they are familiar. Episodic memory of a specific situation's salient features (e.g., cues or indicators, alternative responses, selection criteria, selected alternative, results) can be used as a benchmark for comparisons to a new situation. For example, sessions in which physicians used archival data revealed a tendency to predict or estimate a diagnosis based on analogies to cases they had encountered previously.

When using this method the analyst should request that the domain expert note his or her use of analogies to previous problems. Items of particular interest include:

- Comparison of similarities and differences to the current scenario.
- The types of previous scenarios that are triggered or recalled by the task at hand.
- The identification of the problem's salient features based on experience with other problems.
- Alternative selection criteria that may be stimulated by previous scenarios.

An example of this technique is a session in which the analyst selects a problem in the aircraft emergency procedures domain to investigate problem recognition, categorization, and response. He or she selects a problem involving the co-occurrence of two emergencies. Procedural manuals often spell out what to do in the case of a single emergency, but the solutions may be contraindicated when multiple emergencies occur. The decision maker is presented with the "emergency" and relevant supporting information and is asked to suggest solutions. In particular, he or she is encouraged to note analogies between this problem and similar, previously solved problems. During the ensuing discussion the decision maker's past experiences help to reveal salient features of the current problem.

Difficult Case Analysis

The requirements for any system must address not only functions of the system under normal conditions or with normal cases, but also with difficult, complex, or challenging cases. After process tracing sessions have been conducted to examine the normal process and procedures, sessions should be conducted to look at more complex examples. Findings from these sessions may have a considerable impact on requirements, decision support, and screen design.

The analysis of difficult cases can be handled in a number of ways. One way is to ask experts to select one or two cases that were particularly complex or difficult prior to conducting the session. Using a form such as the one shown in Fig. 10.6, they can characterize the session and provide the analyst with information about the case prior to the session.

For example, emergency medical systems paramedics were asked to identify an incident that stressed the process of responding to emergencies. They selected an incident that involved two locomotives, several hundred passengers, lack of information, and access problems. Prior to the session they were able to provide the analyst with a videotape that described basic features of the incident. The ensuing session explored, step by step, the procedures and decision making that occurred in managing the emergency response to the incident. As a result, analysts were able to identify steps in the basic procedure that were handled very differently in the complex incident, which impacted the functional and communication requirements of the system under development.

Incident/Case Title: AMTRAK Incident
Identifying characteristics or features enhancing difficulty:
- Time of day
- Weather
- Several hundred people involved
- Ingress and egress problems
- Communication overload
Part of the normal process impacted by these features:
- Triage
- Resource management
Sources for familiarization with the incident:
- Videotape

FIG. 10.6. Documenting difficult cases.

An alternate approach is to ask selected decision makers to record (e.g., annotate, audiotape, or videotape) difficult cases as they experience them during a week or other predetermined period of time (Hoffman, 1987). For example, if they troubleshoot the process for making potato chips, they would keep a log (supported by photos when possible) of the line or of problems experienced in the course of a 2-week period (McGraw, 1994b). These notations should include:

- Case description.
- Pertinent products or work artifacts (e.g., reports).
- Solutions considered.
- Factors and constraints managed in determining a solution.
- Reason for action selected.
- Results.
- Why the case was considered difficult.

In a follow-up session the decision maker shares the log with the analyst and together they review each of the difficult cases, the factors contributing to the complexity, and the process used to solve it.

Another alternative is to ask a small panel of experts to evaluate a difficult case or incident and how it was handled. This would involve attempting to explain the following:

- The salient features of the case.
- Key factors and constraints that appeared to be considered.
- Alternatives that could have been accomplished or selected.
- Possible reasons for the approach selected.
- Effectiveness of the results.

Managing the Verbalization and Explanation Process

The amount of information about decision making and solution processes revealed by the decision maker during the session depends in part on the type of verbalization method the analyst chooses to apply. Two often-used methods include concurrent verbalization (i.e., "think aloud") and retrospective verbalization, which Table 10.3 compares.

Concurrent Verbalization

In concurrent verbalization the decision maker solves the problem and thinks aloud or verbalizes actions, considerations, decision points, and strategy. This method immediately reveals what the decision maker believes

TABLE 10.3
Comparing Methods for Creating Protocols
(McGraw & Harbison-Briggs, 1989)

Concurrent Verbalization ("Think Aloud")	Retrospective Verbalization
Decision maker reports on activity as the task or problem is completed	Decision maker completes a task without description prior to discussing his or her activity with the analyst
Requires selection of a task or problem that may be completed in the time allotted to the session	Requires selection of a task or problem that may be completed in the time allotted to the session while still leaving time to allow post-solution descriptions
Can interfere with decision-making and problem-solving processes, which can negatively impact the quality of the information	Requires the decision maker be able to remember what he or she was thinking "after the fact" using the memory aid, which may negatively impact the quality of information
Requires little equipment	Requires recording equipment to produce a memory aid (i.e., tape)

is true about his or her decision-making or problem-solving process. This ongoing commentary benefits the analyst; however, it can be distracting for the performer. Try to solve a complex math problem while discussing how you are doing it, and you will agree that verbalization requires cognitive resources. When decision makers expend cognitive resources on the explanation of behavior, it limits resources available to complete complex mental tasks. Negative results may occur, such as inaccurate or incomplete information. The source of these errors include:

- Omission of steps in the problem-solving or decision-making process.
- Failure to recognize and thus to explain, a step or consideration.
- Interruption of the flow of information or process activity (i.e., "losing one's place").

These negative effects are more predominant when the problem being solved or decision being made does not allow for interruptions. The negative effects are less harmful in situations in which there are natural breaks in the process. As Waldron (1985) explained, negative effects would be greater if we required concurrent verbalization from a pilot who may have to respond quickly to changing conditions. Conversely, interruptions may cause fewer problems in situations such as a mathematics professor solving a difficult problem, because he or she may naturally pause to consider the next step in the solution process.

Retrospective Verbalization

Retrospective verbalization, or cued recall, requires the decision maker to solve the problem or make the decision as he or she normally would—without describing each step (although some people do "self monitor" or talk to themselves as they work). After the decision is made or problem solved, the analyst asks the decision maker to review a memory aid (usually a videotape, but also could include audiotape transcriptions or session notes or graphs the analyst made). The analyst asks the decision maker to talk about what he or she was doing, considering, and why.

This technique is useful if the analyst believes the task of talking about a process while completing it will interfere with the solution or decision process. However, this method is not without problems. Analysts must attend to and record information from the session that may provide cues to performer behavior. As important as the cues themselves is the knowledge of the context in which they occurred. (Unfortunately, unless the analyst is experienced in the domain, he or she might not recognize important features.)

If the verbal report strays from information that seems important, the analyst should guide the decision maker's activity and verbal report by calling attention to relevant behavioral cues and their contexts. Taping these sessions is useful, to provide a memory cue for the analyst to use during protocol analysis sessions.

PRODUCING THE PROTOCOL

The goal of process tracing is to determine the decision-making processes used in the completion of an activity or the solution of a problem. First, the analyst must translate verbal reports and other cues (e.g., tapes) into protocols, or written records. Using this protocol, the analyst can trace the expert's decision-making or problem-solving process from the problem's presentation to its resolution.

Protocols are analyzed to identify the following:

- General problem-solving approach and strategy.
- Decision points.
- Initial hypotheses.
- Apparent impact of attributes of the problem on the problem-solving process.
- Decision rules and heuristics applied.

- Alternatives considered.
- "Tests" used to confirm or deny hypotheses or remove alternatives from consideration.

Translating Sessions to Protocols

One of the outputs of a decision process tracing session is a recorded or annotated history of the session. The analyst translates both verbal and nonverbal session information into the protocol format. The resulting protocol provides a record not only of what the decision maker said, but also cues that may reveal his or her level of confidence, how long he or she searched, or how long the issue was considered before retrieving and/or responding, and so on.

Verbal cues consist of the actual words a decision maker uses and the paralinguistic cues (e.g., "uhmm," tone of voice) that accompany a communication. Analysts can record verbatim information in regular text format, and indicate paralinguistic cues, pauses, etc. within parentheses or brackets.

Nonverbal cues also are laden with information, but some of it (e.g., eye movement) is difficult to monitor and interpret without special equipment. However, some nonverbal cues can be captured by a video camera or analyst/scribe. The following types of information can be recorded in the protocol within parentheses or brackets. In follow-up sessions the analyst can present the tape or describe the context for the behavior and explore what the decision maker was thinking or feeling.

- Pushed away from materials.
- Moved options side-by-side.
- Sighed, grunted, appeared perplexed.
- Scratched or touched head or brow.
- Chewed lip, pencil, etc.
- Squinted.
- Doodled; wrote formulas, etc.

The following guidelines describe the techniques used to reduce a protocol, enabling easier analysis. Although these guidelines are presented as steps, experienced analysts will be able to accomplish all steps concurrently to create a clear, reduced, usable protocol.

1. *Organize and standardize the protocol.* Transcribe or listen to the original words. Identify conjunction points, which indicate the occurrence of multiple activities (mental or physical). Then break original words into separate

TABLE 10.4
Organizing and Standardizing a Protocol Using Additional Space

Decision Maker's Original Statement	Organized Statement
"We'll put this on auto by switching from manual to auto. That's okay now, it's on auto. It'll take a few minutes for the feeder to settle down, so we'll wait."	Put the potato chip cooker on auto by switching it from manual to auto Confirm it is on auto Wait for it to stabilize (1 minute)

activities by adding slash marks (/), starting a new line, or tabbing to separate activities by space. If time is important to the decision process, add notes to indicate elapsed time. Table 10.4 provides an example of organizing and standardizing a protocol using additional space.

2. *Reduce and refine the protocol.* Continue to manipulate the protocol by eliminating extraneous vocabulary. For example, in the example shown previously, the analyst would eliminate words like "Well," "that's okay now," and so on. Next, reduce synonyms or acronyms the decision maker used to a restricted set of common words and phrases (i.e., terms stored in a project glossary and used consistently by all of the project's team members). This will be useful later in the project in defining objects and their functionality. Finally, eliminate redundancies (where the decision maker repeats a statement) and resolve ambiguities. Ambiguities include incidences in which the decision maker used an antecedent to refer to a pronoun, or a pronoun to refer to a noun. Table 10.5 illustrates how this reduction activity affects the protocol.

3. *Standardize the format of the protocol.* The format should be standardized across the program to enable all analysts, designers, and developers to understand its content. It can be set up as a template that can be easily used, stored, and retrieved. Figure 10.7 illustrates a standard format for a protocol. This form can be modified as needed by the program. For example, the project may decide to add a column for Timing, in which elapsed time could be recorded.

TABLE 10.5
Reducing and Refining the Presentation

Organized Statement	Reduced Statement
Put the potato chip cooker on auto by switching it from manual to auto Confirm it is on auto Wait for it to stabilize (1 minute)	Put the potato chip cooker on auto by switching cooker from manual to auto Confirm potato chip cooker is on auto Wait for potato chip cooker to stabilize (1 minute)

| Session # _____ Analyst: _____ Interviewee: _____ |
| Tape Identifier: _____ Topic: _____ |

LINE	CONTENT	STATEMENTS
001		
002		
003		
004		
005		
006		
007		
008		
009		
010		
011		
012		

FIG. 10.7. Sample format for a protocol template.

The Line section shows line numbers down the left-hand side. Line numbers enable the analyst to quickly locate a specific piece of information. This makes it easy to work with a decision maker and point quickly to the line in question. The Content section represents the reduced, standardized protocol. The Statements area on the right is used to extract specific information in a structured English format and present it to eliminate subsequent searching of the transcript by each new reader. Statements may be structured in a decision format or may represent a comment or summary of the information presented in the description. A decision format (e.g., If/Then/Else) should be used when two or more actions can be taken depending on the value for a specific condition. The Statements area may be completed after the protocol has been analyzed, or while working with the decision maker during the analysis of the protocol.

ANALYZING THE PROTOCOL

A thorough analysis of the protocol involves a series of tasks, including:

- Reviewing the tape and annotations from the process tracing session.
- Conducting the protocol analysis session with the original decision maker.
- Identifying the basic alternatives the decision maker seemed to consider at specific decision points and the attributes, aspects, and attractiveness of each alternative (Waldron, 1985).
- Identifying reasoning heuristics used by the decision maker.

The sections that follow describe these activities in more detail.

Conducting the Protocol Analysis Session

The analysis process often requires a protocol analysis session with the decision maker that should occur as soon as possible after the decision process session has been conducted. The analyst should review the tape or notes from the process tracing session and produce at least a partial protocol prior to the protocol analysis session. He or she can use the Statements area of the protocol template to identify questions that should be asked about a particular process step, heuristics revealed that require validation, and so on. The analyst will use the protocol and the tape from the process tracing session to facilitate the protocol analysis session.

Protocol analysis sessions may be simple follow-up interviews with the decision maker who participated in the process tracing session. Alternatively, they may involve group discussions, in which the original decision maker and one or two peers participate.

The most common protocol analysis session is an interview format with retrospective verbalization. In this type of session, the analyst plays back the tape from the process tracing session and asks the decision maker to provide a verbal report of actions, considerations, decision points, strategy, etc. When the analyst needs to interject questions, he or she stops the tape and uses questioning techniques (see chapter 7) to elicit the required information within the context of the process.

In the discussion format session, the analyst asks two or more decision makers to work together to review and verbally annotate a taped problem and solution. Their actions, explanations, and other communications, including areas of conflict and varying perspectives, are noted on the protocol form. A primary benefit of the discussion format is that the analyst is able to benefit from the expertise of more than one decision maker, thus moderating the possible effects of bias that may exist in the single-decision-maker interview format.

Analyzing Decision-Making Factors

The identification and analysis of decision-making factors includes an investigation of alternatives, their attractiveness and attributes, and reasoning heuristics.

Identifying Alternatives

At the simplest level, decision making can be classified according to the major alternatives the decision maker identifies. Alternatives represent possible actions on the decision maker's part, and provide information relating to probable outcomes as a consequence of selecting each alternative.

If a decision maker is monitoring gauge readings to troubleshoot a faulty system function, he or she will have several alternative actions for a situation in which one of the readings approaches an "out-of-upper-range" reading. One alternative may be to consult readings of other gauges prior to taking action. Another may be to respond by physically switching off a system or manually decreasing an input to the subsystem. Each alternative may have different outcomes that the analyst should explore in terms of its costs and potential benefits.

In some systems (e.g., diagnostic, troubleshooting, etc.) the focus is on the hypotheses the decision maker forms based on the data and patterns he or she recognizes. These hypotheses immediately enable the decision maker to limit his or her search through the problem space, and plan an approach that confirms or rejects the hypotheses. Consequently, analysts must be able to identify these hypotheses and the alternative actions associated with each. They must also document the "tests" the decision maker applies to confirm or deny each hypothesis and select the appropriate alternative and response. Figure 10.8 illustrates a partial transcript from a process tracing session (Lancaster, 1989). Each hypothesis is marked (H1 for hypothesis #1, etc.). The test used to confirm or deny a hypothesis is also marked (T1 indicates the test used to confirm or deny Hypothesis 1, etc.).

Psychologists have always been curious about how people compare alternatives or hypotheses to make decisions and solve problems. The results of more than 2 decades of research in decision making has helped to reveal the reasoning normally applied. Decision makers compare alternatives by analyzing each alternative's attributes, aspects, and attractiveness.

Analyzing Attributes of Alternatives

Attributes are the separate components or dimensions that define an alternative. For example, if we are attempting to buy a car we will analyze the

Well, I guess I could see if it will start.	
It could be a bad starter.	H1
Or it might not be getting fuel.	H2
or electrical connections.	H3
(Attempts to start.) No, it just grinds.	T1
Now I will open the hood.	T2
I am checking for wiring problems because I think there might be a bad connection.	T3
Yes, the cap is off, so there wasn't a good hook to the spark plug.	H3

FIG. 10.8. Identifying hypotheses to document alternatives and tests to confirm or deny each.

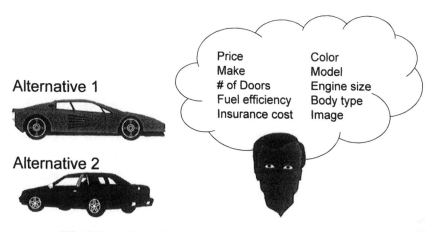

FIG. 10.9. Attributes for the comparison of car purchase alternatives.

problem by devising alternatives that are defined or expressed by attributes, such as those shown in Fig. 10.9.

Analyzing Aspects of Alternatives

Just as each alternative can be expressed more specifically by its attributes, so do aspects further define attributes. Aspects are the "values" that we assign to a specific attribute. For example, in Fig. 10.9, ten attributes will be used to compare alternatives. Each attribute can have a number of aspects that might be considered. For example, *Price* has a value or aspect that varies for each alternative. The aspect of Price for Alternative #1 might be $30,000, while the aspect of Price for Alternative #2 might be $17,500. Table 10.6 presents examples of aspects for other attributes.

TABLE 10.6
Aspects for Selected Car Attributes

Attribute	Aspect or Value
Color	Candy apple red Navy blue
Body type	Sports car Sedan
Make	Ford Mercedes Chrysler
Image	"Adventurer" "Family person" "Business minded"

Analyzing Attractiveness of Alternatives

The analysis of an alternative's overall attractiveness provides another dimension analysts can explore by reviewing the protocols and probing the selection of an alternative approach with the decision maker. Attractiveness is the psychological weight or strength of an aspect, or may reflect the frequency of its selection. Attractiveness should reflect the decision maker's analysis of an alternative's costs versus benefits. It should be possible to represent attractiveness with a single value, which can then be used to compare alternatives on a single aspect (Waldron, 1985). After the decision maker has compared alternatives and their aspects, he or she can rank them based on their overall attractiveness.

Identifying Reasoning Heuristics

Protocols can also be analyzed by examining the overall reasoning approach and heuristics a decision maker seems to be using. Research in the area of human decision making (Lee, 1971) has identified naturally recurring reasoning approaches, including those that follow. Each of these is explained in more detail in subsequent sections:

- Dominance approach.
- Conjunctive approach.
- Disjunctive approach.
- Lexicographic approach.
- Elimination by aspects approach.
- Maximum attribute attractiveness approach.

Dominance Approach. The purpose of the dominance approach is to select the superior alternative. The decision maker compares the attributes of each alternative with attributes of each of the other alternatives (Lee, 1971). The superior alternative is the one that is better than all other alternatives on at least one attribute, and not worse than any alternative on all other attributes.

Conjunctive Approach. Using the conjunctive approach, the decision maker selects attributes that are critical to an alternative. The decision maker then specifies a set of criterion values for each selected attribute. Any alternative that has even one attribute that does not meet the criterion value is discarded (Coombs, 1964; Dawes, 1964; Svenson, 1979).

Disjunctive Approach. This approach also requires the decision maker to identify each attribute that is critical to an alternative and specify a set of critical criterion values for each attribute. The decision maker identifies alternatives with at least one attribute that *exceeds* the criterion, then selects from among these alternatives (Coombs, 1964; Dawes, 1964; Svenson, 1979).

Lexicographic Approach. The lexicographic approach is a simple ranking of attributes for each alternative according to their perceived importance (Fishburn, 1974). The decision maker ranks each attribute of each alternative, listing the most important attribute first. He or she then compares the most important (e.g., top-ranked) attribute of each alternative and selects the alternative that is most attractive based on the most important attribute.

Maximum Attribute Attractiveness Approach. This approach enables the identification of the *most* attractive alternative based on the sum of all values for an alternative's attributes (Svenson, 1979). The decision maker sets criterion values for the attributes of each alternative, then ranks the values for each attribute against competing attribute values. Ranking can be as simple as "equal to," "better than," or "worse than." The best alternative is the one with the most attractive attributes.

Elimination by Aspects Approach. This approach combines features of the lexicographic and conjunctive approaches (Tversky, 1972). The decision maker identifies each attribute that is critical to each alternative and specifies a set of criterion values for each critical attribute. He or she then ranks the attributes of each alternative, listing the most important attribute first, and compares the values of the ranked attributes. The decision maker eliminates from consideration any alternative that does not equal or exceed the conjunctive criteria for its attributes. Finally, the decision maker repeats the procedure with attributes that fall lower in the lexicographic order to eliminate the alternatives whose aspect values do not equal or exceed the conjunctive criteria.

TYPICAL OUTPUTS OF PROCESS TRACING SESSIONS

In addition to the protocol, other types of output are beneficial in communicating the decision-related information elicited from these sessions. Each of the following is a useful tool in documenting decision related information. Subsequent sections describe them in more depth.

- Decision trees.
- Decision/action diagrams.
- Decision tables.

Decision Trees

A decision tree is a diagram that presents conditions and actions sequentially. Its purpose is to enable analysts to describe the order in which conditions should be considered, and to depict the relationship of each condition to its permissible actions. These diagrams take the shape of fans, or trees, as shown in Fig. 10.10. The root of the tree depicts the point at which the decision sequence begins. The tree's branches represent conditions, and indicate that a decision should be made about which condition exists before the next path is followed. The right side of the tree lists actions to be taken, depending on the sequence of conditions followed (Senn, 1989). The branch that is followed depends on the conditions that exist and the decision that needs to be made. As you make a series of decisions, you progress from left to right along a particular branch. After each decision point, the next set of decisions to be considered is depicted.

Decision trees are particularly useful when the analyst needs to depict the sequence of decisions. Decision trees are developed after a work process or task analysis has been conducted and thus, the flow of actions and data through business processes or tasks has already been identified and defined. This greatly simplifies the task of constructing the decision tree. Decision trees may not always be the best representation of problem-solving activity, however. They quickly become overly complex and burdensome if they represent a complex system or a problem with many sequences of steps or combinations of conditions.

A variation of the decision tree is the "fishbone" diagram, so called because it is shaped like the skeleton of a fish. This format is particularly useful in domains in which the decision maker might have to consider several hypotheses and conduct several tests to identify the problem. When creating these diagrams, analysts should represent the more common hypotheses near the "head" of the diagram, and uncommon or frequently selected hypotheses near the "tail."

FIG. 10.10. Sample decision tree.

Decision/Action Diagrams

Decision/action diagrams depict the progress through a system in terms of the decisions and actions that must be performed (Kirwan & Ainsworth, 1992). Using standard notation for diagrams, decision points are represented as diagrams. Actions are denoted by rectangles. Possible outcomes are labeled on the exit lines. Questions may be formatted in a yes/no or multiple choice format. Figure 10.11 illustrates a simplified decision/action diagram for the process of identifying aircraft and assigning runways.

Decision Tables

Decision tables represent *all* the conditions that must be true for an action to be taken, rather than one condition at a time. A decision table is a matrix of rows and columns that depict conditions, actions, and decision rules that state what procedure to follow when certain conditions exist (Senn, 1989). Simple decision tables are comprised of conditions, actions, and notations that detail whether the action should be taken in each condition. Conditions include *condition statements*, which identify relevant conditions, and *condition entries*, which tell the value that applies for a particular condition. Actions include *action statements*, which list the set of steps that can be taken when a certain condition occurs, and *entries*, which show what specific actions in the set to take when a condition or combination of conditions are true.

Extended decision tables provide even more information. Instead of marks or notations (e.g., an "X"), the table presents words or phrases. Figure 10.12 illustrates an extended decision table that might be used by

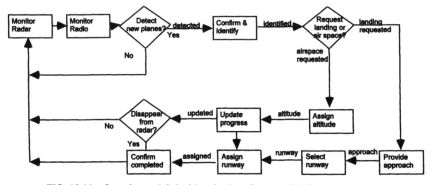

FIG. 10.11. Sample partial decision/action diagram (McGraw & Harbison-Briggs, *Knowledge Acquisition: Principles and Guidelines*, 1989). Reprinted by permission.

Conditions	Decision Rule 1	Decision Rule 2	Decision Rule 3	Decision Rule 4	Decision Rule 5
Patient's Age	Patients <10 yrs old	Patients >50 yrs old	Patients between 10 and 50 yrs old	Any	Any
Depth of Burn	2nd or 3rd degree	2nd or 3rd degree	2nd or 3rd degree	Any	1st degree
Body Surface Area (BSA)	<10% of BSA	<10% of BSA	>20% BSA	Face, hands, feet, or perineum	Any
Actions	Refer to Burn Center	Refer to Burn Center	Refer to Burn Center	Refer to Burn Center	Nearest medical facility

FIG. 10.12. Sample extended decision table.

a paramedic for making decisions relative to the treatment of a burn patient.

To complete an extended entry decision table:

1. To identify the conditions, select the most relevant factors to be considered in making a decision. Each condition should have the potential to either occur or not occur, or to have variables associated with their occurrence.

2. To identify the action entries, determine the most feasible considerations, variables, or activities for each condition. These short phrases will signal the correct action statements.

3. Analyze the protocol and follow-up session information or work with a decision maker to identify the combinations of conditions that are possible.

4. Fill in condition rows with action entries. Action entries (e.g., <10% of BSA) tell the user what information should be used to decide.

5. Fill in the action that is appropriate for each condition. For each rule, enter a short phrase (i.e., "Refer to Burn Center") in the action statement section.

6. Ask an expert decision maker to review the decision table to ensure its accuracy and completeness, but also to identify any contradictions or redundancies. If a contradiction is found, some of the information may be inaccurate and must be corrected. However, the analyst should not discount the fact that different decision makers may come to the same conclusion using different rules. If redundancy occurs (i.e., multiple decision rules are identical except for one condition and the actions for the rules are identical), combine rules.

SPECIAL CONSIDERATIONS OF PROCESS TRACING AND VERBAL REPORTING

This set of techniques is not without critics who argue that the information they yield is not as valid as desired and that the yield amount is not worth the time required to produce the output.

We agree that this set of techniques reflects some of the failings of expert decision making—particularly biases. However, we argue that biases themselves are rules of thumb that need to be considered. Even experts often do not make optimum decisions because they rely on heuristics or rules of thumb rather than on statistical probabilities and formalisms. Heuristics can lead to satisfactory solutions, but some research has shown that this is not always the case (Simon, 1981). Heuristics also may be associated with biases. These biases are often triggered by the context of the decision making. As Kahneman and Tversky (1982) noted, "the actual reasoning process appears to be schema-bound or content-bound so that different operations or inferential rules are available in different contexts" (p. 499).

Another consideration is that the analyst must be ever diligent in focusing on the elicitation of the decision-making knowledge, which is semantic in nature. Do not fall into the trap of focusing more on the expert's *explanations* (which are declarative in nature). It is also reasonable to consider the types of knowledge elicited during the process tracing and protocol analysis session. The use of process tracing and protocol analysis is often defended using the argument that we must understand the heuristics or rules of thumb experts use in solving problems and making decisions. This type of knowledge is semantic, or deep knowledge. As it is applied and as the decision maker follows ingrained procedures to complete tasks, analysts can observe the procedures that are conducted, but not the rules used. To elicit the rules, analysts either ask the decision maker to think aloud and describe what he or she is considering, or explain behavior using a cue such as a tape. Unfortunately, some analysts often focus more on the descriptions of activity, which yield declarative or surface level knowledge, rather than spending the time required to test the heuristics that are being stated.

Analysts also must recognize that decision makers asked to explain their activity will respond using one of two strategies to generate explanations. If the decision maker's search procedures enable them to access appropriate strategies that they can verbalize, the process is effective (Nisbett & Wilson, 1977). In this case the explanations are likely to be a reasonable representation of the heuristics applied. However, if the decision maker's search procedures do *not* yield appropriate strategies, he or she may attempt to create a *plausible* (but not necessarily accurate) explanation. Expert decision makers construct explanations by observing information (e.g., situational cues) and their own behavior and using domain-general skills to infer rules that could help them rationalize the behavior (Gordon, 1992). In this case the explanations are probably faulty and are not generalizable representations of the heuristics applied.

Diligent analysts can learn to apply these process tracing and protocol analysis techniques and manage the special considerations required. First, we must recognize that bias exists and ask probing questions to investigate the possible presence of bias in responses. We must carefully select problems for use in these sessions and work with experts to identify possible ranges of response and to review responses from single experts. We may present the same problem to a small group or panel of experts to guard against the bias that may occur when using only one expert. Additionally, we use other techniques, such as observation and task analysis, to help document and confirm frequencies of tasks, decisions, and preferred alternatives.

Secondly, analysts must focus more on problem setting, observing the activities involved in decision making, and probing for specific information related to reasoning process. Inexperienced analysts who let the protocol analysis session become a simple interview, or worse yet, a lecture from the expert, elicit only declarative information that fails to provide the depth required.

Finally, analysts must seek direct verbalizations rather than rationalizations. Analysts must introduce the sessions in a way that encourages experts to confess when they don't know why or need further information to provide acceptable rules. Additionally, analysts must carefully observe expert decision makers for signs (nonverbal and verbal) that they are using an ineffective strategy to generate explanations, building their own rational mental model as they go.

Chapter 11

Conducting and Analyzing Group Sessions

The elicitation techniques presented in previous chapters of this text focused on an analyst working with an individual performer, or a series of performers, to accomplish a specific set of goals. Although multiple performers would be consulted during the use of these techniques for requirements, analysis, and design activities, to this point we have discussed approaching them individually.

Analysts who find sessions with a single performer challenging may find group work daunting. When working with groups, analysts must not only be proficient in the technique being used, but also in group management, facilitation, and understanding group dynamics. Furthermore, these sessions demand more planning and follow-up, and are more time consuming to translate and transcribe.

In spite of the difficulties associated with group sessions, however, there is much to be gained. Sessions with groups help reveal multiple perspectives and lines of reasoning, producing better collective information and more creative solutions than a single source. Additionally, a group can benefit the project politically as the group coalesces and begins to "buy in" to the goals of the project.

This chapter first answers the question "Why work with groups?" to help analysts who may need to convince others (or themselves!) of the need to do so. Next, it presents a core set of group techniques, guidelines for their use, and examples of how they can be varied. The chapter also presents ideas for using computer applications, such as mind mapping and video conferencing technology, that can be used to support group sessions. Finally, the chapter concludes with a discussion of some of the challenges associated with group sessions.

The purpose of this chapter is to present information that will enhance the analyst's use of group elicitation and analysis techniques with multiple performers. Specifically, it enables analysts to meet these goals:

- Be able to defend the use of established group techniques during requirements elicitation and refinement, prototype component review and refinement, scenario development, and document review.
- Determine when to use brainstorming, consensus decision making, nominal group technique, and focus groups.
- Use computer and video conferencing technologies to support the use of group techniques.
- Confidently facilitate group sessions.

WHY GET INVOLVED WITH GROUPS?

Many analysts are reluctant to become involved with the use of group elicitation techniques or with the facilitation of a group. However, analysts must consult groups of performers and/or management during the development of requirements and prototype components because of the following issues:

- Complex domains and expertise.
- The need to obtain consensus and support.
- Benefits of group sessions.

Complex Domains and Expertise

Complex, real-life problems are seldom so simple that they can be solved by eliciting information from a single performer. Even when multiple performers are consulted, analysts may find it difficult to compile diverse, often conflicting information.

Specialized Expertise

Perhaps the most common reason to use groups is when the system under definition (a) spans multiple functional areas, (b) will be used by people holding different job positions, or (c) will be used by people in the same position in different geographic regions. When the project spans multiple functional areas or job positions, the domain can usually be segmented into these area. People in each of these areas might require specialized expertise to perform their job. Analysts would work individually with target performers in each area, but could enhance the project's success by getting the "big picture" of a process through a group session. For example, work on a project at an integrated circuit manufacturing facility revealed these performer groups: machine vendor engineer, equipment engineer, process engineer, senior equipment technician, and quality engineer. Each represented a different job function. Pulling them together to review system requirements and prototypes provided the overall picture of the process and enhanced the system's completeness and functionality.

Furthermore, the same job position may differ based on geographic region. Performers on the west coast perform under different conditions, constraints, and variations in the application of heuristics from those working on the east coast of the United States. Pulling together a set of performers from all over the country can ensure that the analyst takes advantage of the specialized expertise and constraints that exist in each locale.

Varying Perspectives Based on Experience

It is simplistic to assume that any one person has the knowledge, skills, and ability to provide enough information and data to build project components that meet the requirements of complex domains. Working with a small set of individuals may not provide diverse enough perspectives on issues that are particularly complex. Any one individual looks at and reacts to a scenario, question, or problem to solve from his or her own perspective, based on experience and expertise. That perspective is probably valid, but may not be the *only* valid perspective. Working with individuals clarifies the issues—working with a group refines the issues to support varying perspectives and ways of interpreting, approaching, and solving domain problems. For example, analysts for a project at Ford Motor Company's robotics center noted that "a person may be an expert, but that doesn't mean he is always right" (Hunter, 1985).

Gaining Consensus and Building Cohesiveness

One of the most potent reasons to work with groups during requirements elicitation, scenario analysis, prototype component review and refinement, and document reviews is to build consensus. Consensus includes gaining the group's "buy-in" and support, and agreeing on approaches.

Group Buy-in

Building consensus and gaining group "buy-in" for the visual and functional requirements of a project is important for two reasons. First, project quality and completeness is positively affected by the "tweaking" that occurs as a competent group offers suggestions and support. Second, as the group manipulates and comes to accept the ideas presented, they become more supportive of the project because it reflects not only ideas they support, but ideas colleagues also support. This produces a valuable "community of support" and a synthesis of expertise.

Cohesiveness

Once an effective group of performers has been identified, and has worked together successfully, analysts may want to consider asking the group to participate in follow-up sessions and/or reviews. As a group works together, comes to understand project goals, and completes sessions, a cohesiveness often develops that can benefit software projects in terms of both quality and politics.

The positive political ramifications that result from gaining buy-in and building cohesion can greatly aid the project and enhance opportunities for its success. For example, we recall a focus group session held with potential users of a new system. The group worked surprisingly well together, in spite of the fact that participants represented all parts of a complex project and various levels of rank and status. The group session yielded not only good ideas, but also political support for the user-centered project. When the project was stalled due to resistance to change at one level of the organization, the focus group participants and their management worked behind the scenes to rekindle the support required to move ahead.

Additional Benefits of Group Sessions

Additional benefits of group sessions include the importance of synergy, quality of results, and the ability to identify (and reject) poor solutions.

Synergy

Synergy is the creation of a whole that is greater than the sum of its parts. It enables groups to accomplish more than the total of their members' individual capabilities. Part of this is due to the fact that groups tend to take more risks than individuals working alone. Synergy also is believed to help well-managed group sessions yield more creative, innovative solutions than an individual does (Schermerhorn, Hunt, & Osborn, 1991).

Quality of Results

On the average, small group judgment tends to be more accurate and better than individual judgment (Eisenson, Auer, & Irwin, 1963), or at least as good as the judgment of an average individual. In fact, a group of performers can provide superior results when the problem can be partitioned into related subproblems and each member's skills matched with a particular subproblem (Steiner, 1972). Thus, the group is working and making decisions together, but when a question arises the individual with the appropriate skill set is given a "louder" voice.

Identifying New Opportunities

In addition to providing multiple ways of looking at a problem in a complex domain, group sessions may also result in the identification of new, previously unconsidered opportunities. It is often the case that analysts have met with four or five performers and identified problems and wish lists a new system could solve, brought it to a group to obtain consensus, and had group members "springboard" off of the initial ideas. New opportunities identified by group members may take the form of refinements to initial requirements, scenarios, or desired functionality, or may reflect new ways of thinking about the problem.

Identifying and Rejecting Poor Solutions

One of the most positive outcomes of a group session is a group's ability to recognize and reject incorrect, invalid, or unrealistic solutions and suggestions. An individual working with an analyst may be reluctant to provide negative feedback, or simply may not see possible flaws. A group is able to critique a solution, approach, or prototype component from multiple perspectives.

TECHNIQUES FOR USE WITH GROUPS

Analysts may modify techniques used with individual performers, a pair or small group of performers, or use special techniques designed for use with larger groups.

Modifying Techniques

Many techniques previously discussed in this book can be used effectively with a small- to medium-size group (i.e., three to five people). The use of these techniques with small groups is most feasible when the goal is to review and refine a product or output of a session. For example, small groups are particularly effective in these activities:

- Participatory design sessions to evaluate storyboards for a proof of concept for the system.

- Participatory design sessions to evaluate screens being developed for the system.
- Scenario development to define what the "to be" system will do in specific situations.
- Documentation review during requirements and design phases.

One key difference in using techniques with groups is the responsibility of the analyst. Working with pairs or small groups requires the analyst to facilitate, or help enable, the interaction among participants while moving the session toward its goals. Techniques that are most conducive to modification include work process or task analysis walkthroughs, decision process tracing, and scenario analysis. Each of these is focused on something such as a process, task, problem, or scenario.

Work Process and Task Analysis Walkthroughs

Bring a small group of participants together to review a work process diagram, or a task analysis already developed. Each participant receives a copy of the diagram or analysis, which is also projected on a screen. The analyst focuses the discussion on each block in the process or on each task in the set and asks questions to ensure its accuracy, completeness, and special considerations. Or the analyst can ask the group to work together to demonstrate or act out the completion of a task or work process not yet represented in graphic form.

Decision Process Tracing

Decision process tracing and protocol analysis are paired techniques that enable analysts to trace and study a performer's decision-making processes (see chapter 10). By modifying the way a problem is presented, analysts can effectively use decision process tracing with small groups of performers. Some researchers (Wielinga & Breuker, 1985) contended that a pair of performers or a small group provides superior solutions because it reveals the decision-making process more clearly. Observing a small group solving a problem together enables the analyst to observe and isolate the considerations, guidelines, constraints, and points of conflict that apply.

Special Techniques for Groups

Special group techniques that can be used for domain analysis and requirements elicitation as well as for document and design reviews include brainstorming, consensus decision making, and the nominal group technique. How and when to use these techniques, along with variations of their use, is presented in the next sections.

Using Goal-Oriented Brainstorming

Brainstorming (Osborn, 1953) originated from concerns by business executives that lower-level managers were too quick to parrot conventional wisdom of their superiors. This approach was politically safe, however, it didn't result in the best strategic planning and decision making. Brainstorming is a technique that promotes the identification of a number of responses to an issue prior to any analysis of the ideas or decision-making.

Traditional Brainstorming. Analysts can use the traditional brainstorming approach to generate ideas such as requirements, demonstration scenarios, and so on. Brainstorming can help participants and analysts break loose from the obvious, conventional solutions to complex issues. Performers frequently get into a rut when asked to consider new approaches—the tasks they perform everyday and the way they have always made decisions biases them to these comfortable solutions.

Brainstorming is designed to stimulate thinking and generate ideas to avoid being sidetracked to a quick acceptance of the ordinary and conventional. This technique can be very effective when the analyst needs to prohibit immediate criticism and reduce discussion-inhibiting comments from performers.

"Shout and Post." The "shout and post" method of brainstorming represents its most common, traditional use. The analyst posts a trigger question or stimulus statement and directs the group to call out ideas, solutions, alternatives, etc. as they come to mind. This method results in a facilitated free-for-all in which members can shout out their own ideas or use a previously presented idea as a springboard to a modified version of that idea. Figure 11.1 illustrates this technique.

The shout-and-post technique is effective when the status and power of group participants is equal, or when the organizational culture is such that participants are used to working cooperatively, across division lines and levels to solve goals. To conduct this type of brainstorming session, follow these steps:

1. Introduce the brainstorming session with a brief statement of the session's goals, major considerations, and rules for participant interaction.
2. Present a carefully crafted problem or question to consider, or topic to be brainstormed.
3. Invite participants to call out their ideas as rapidly as they can. Allow them to speak whenever there is an opening in the communication.

FIG. 11.1. Traditional "shout-and-post" brainstorming.

4. A scribe records all ideas on an overhead transparency, whiteboard, flip chart pad, and/or computer.

5. Continue brainstorming until the analyst notices a reduction in the rate of idea presentation. It is extremely rare that a high level of participation will be sustained for more than 10 minutes.

6. The analyst presents to the group the ideas that were introduced, to enable participants to validate their ideas. Only the participant who presented an idea can modify it.

Round Robin. Round robin brainstorming helps ensure that each participant has equal opportunity to share ideas. In this method of brainstorming the analyst uses verbal and nonverbal communication to solicit ideas from each participant in turn. To conduct round robin brainstorming, follow these steps:

1. Introduce the brainstorming session with a brief statement of the session's goals, major considerations, and rules for participants.

2. Present a carefully crafted problem or question to consider, or topic to be brainstormed.

3. The analyst facilitates the session so that each participant is ensured a turn. The analyst uses eye contact and body language to prompt each participant in turn, calling on them if necessary to stimulate input.

4. Each participant presents one idea during a turn or may pass if he or she does not have a contribution.

5. A scribe records all ideas on an overhead transparency, whiteboard, flip chart pad, and/or computer.
6. The analyst calls for multiple rounds until no participant has any other comments to add during his or her turn, or until there are enough ideas with which to work.
7. The analyst reads back to the group the ideas that were introduced and provides a time during which participants can validate their ideas. Only the participant who presented an idea can modify it.

Brainstorming Problem Tree. Another variation on "traditional" brainstorming that can be used to motivate group participants is to use a problem tree.[1] The analyst identifies, or leads the group to identify, a set of problems that must be solved for a particular approach, solution, or system requirement to work. Each problem is written on a different "limb" of a tree graphic. Figure 11.2 illustrates a sample problem tree.

The analyst uses the tree as a stimulus by focusing the group on one limb at a time. The group uses either the shout-and-post or round robin technique to identify as many ways as possible to solve each problem. The analyst/facilitator or scribe documents each set of ideas associated with a problem and posts the resulting set on the wall. Brainstorming activity continues until each limb has been addressed. After solution ideas have been generated, the analyst can plan a consensus decision-making or nominal group session at another time to examine the pros and cons of each solution, and to rank them.

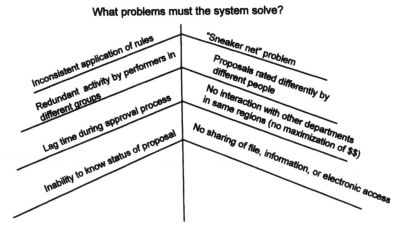

FIG. 11.2. Sample Problem Tree.

---

[1]This technique is similar to the Toothache Tree described by Michalko (1991).

Using Consensus Decision Making

Consensus decision making can be used by itself or as a follow-up to brainstorming sessions. In contrast to brainstorming, in which the focus is on the quantity of the responses, consensus decision making focuses on finding the best solution to a problem, set of requirements, or decision.

Consensus decision making is useful when there are obvious differences in the impact of options or ideas being considered, or requirements being selected. For example, all options and ideas identified during brainstorming sessions are not equally good and easy to implement, nor do they have the same cost/benefit ratio. Consensus decision making is also valuable when analysts need to expose conflict—sometimes it is valuable to recognize that the group can't select one "best" solution or set of issues.

Traditional Consensus Decision Making. Traditional consensus decision making involves presenting a problem and its potential solution, or a requirement and various ways to meet it, to the group and encouraging each participant to vote on alternatives. Voting occurs in rounds, which helps provide a means of gaining commitment to the final solution. The analyst sets the number of rounds based on the number of alternatives or options with which the group will be working. This unique method of voting ensures that each participant has the opportunity to make his or her views known publicly, regardless of rank or status. As voting occurs, the analyst or scribe totals the votes for each option and eliminates options receiving no votes.

General Guidelines for Consensus Decision Making. Consensus decision making is effective *only* if each participant feels his or her views and opinions have been heard. It is also vital that each participant have a commitment to the group's selection(s) even though he or she may have some reservations (Cragan & Wright, 1980).

The analyst follows these guidelines to conduct a consensus decision-making session:

1. Introduce the session with a brief explanation of consensus decision making, the session's goals, major considerations, and rules for participant interaction.

2. Present the problem being considered and possible solutions that have been compiled previously (i.e., in a brainstorming session, by users or other performers, etc.).

3. The analyst announces the number of rounds to be considered based on the number of solutions. For example, three major approaches or solutions would result in three rounds.

4. In the first of the three rounds, each participant would receive three votes. The analyst calls on each participant, who tells the analyst on which options the votes should be placed. Participants may place all votes on one option or split them however they wish.

5. At the end of the first round the analyst reviews the results of the voting and deletes any option receiving less than some agreed-upon number of votes (normally zero, but could be set to another number if the group prefers).

6. If the list of options is large or the voting reflects great variance, the analyst calls for a second round. During the next round each participant receives only two votes (or one less than first allotted) and can place no more than one vote beside any one option. Options with less than a predetermined number of votes are deleted and further discussion can occur as needed.

7. Rounds continue until only two options remain or one option emerges as a unanimous choice.

8. The analyst leads a discussion of the preferred option(s) and asks questions such as:

"Does everyone understand each option?"

"Can any of these options be combined?"

9. Finally, the analyst seeks consensus by asking:

"Does anyone feel you cannot commit to this option or approach?"

"Does everyone feel comfortable with this solution?"

Those who wish to do so can compare a preferred solution with the one selected. This last review of alternatives may reveal solutions that modify the selected option.

Using the Nominal Group Technique

The nominal group technique (Huseman, 1973) is a problem-solving procedure that reduces negative effects such as nonparticipation and conflicts that may be triggered by face-to-face interaction among members of a group.

This technique is useful if characteristics of the group appear to threaten the effectiveness of "normal" consensus decision making. For example, if participants are heterogeneous with regard to rank or status, the analyst might be concerned about the potential for less participation and honesty in open group communication. The nominal group technique enables the participants to become a "group" in name only, as Fig. 11.3 illustrates. Thus, group members are together, but are allowed to function independently and anonymously.

The nominal group technique is often used when a group needs to compare, select, or rank solutions or advantages and disadvantages of ideas.

FIG. 11.3. The nominal group concept—a group in name only.

However, it is also useful in comparing demonstration scenario approaches, scenarios on which to base a prototype, prototype functionality, or a design approach.

Guidelines for Nominal Group Technique. For the nominal group technique to be effective, specific guidelines must be followed, regardless of exactly how they are implemented. Obviously, each participant's anonymity must be ensured. Research (Van de Ven & Delbecq, 1971) has shown that the technique works best when the purpose of the meeting or session is clearly specified and limited to a single goal (i.e., problem solving *or* problem identification *or* refining ideas). Additionally, preparation is essential if the meeting is to go smoothly. This includes collecting (and compiling for review) ideas the facilitator can distribute to participants prior to the meeting, which gives participants an opportunity to review them. It is also useful to ask participants to write additions or modifications next to the ideas on the list and to bring them to the meeting for sharing. Finally, the analyst must provide discussion time for each item on the list *prior* to voting. This ensures that each participant can speak for an idea and attempt to influence other members prior to the vote. The discussion should be limited to a presentation of the pros and cons of an item, and modifications or clarifications of an item—no arguments allowed.

Conducting a nominal group technique session requires the facilitator to work through the steps that follow (McGraw & Harbison-Briggs, 1989; Frank, 1982):

1. The analyst or facilitator determines the type of nominal group session that will be held and communicates its purpose, goals, guidelines, and suggestions for preparation to the participants. For example, the session may be a problem identification session, a problem-solving session, or an idea-, document-, or prototype-refinement session. Each session will require at *least* one hour of active meeting time.

2. When the group convenes, the facilitator defines the nominal group technique and describes how the session will be run. The facilitator makes clear how the session participants should act during the discussion segment (i.e., no arguments).

3. A manageable set of target ideas is presented on a flip chart or projected transparency. For example, ideas to solve a problem may have already been generated during a brainstorming session. Participants are asked to take an index card that represents each idea, and to list its advantages and disadvantages on the card. They complete this task anonymously, and without any discussion. Each participant hands his or her completed cards to the facilitator.

4. Without comment, the facilitator reviews the cards and compiles a list from the items written on each card. For example, the facilitator draws a line down a sheet of flip chart paper and writes all the advantages of an idea on one side of the line, and all the disadvantages on the other side. The facilitator may eliminate only those ideas that are redundant and overlap each other, and may reword only when it is necessary to clarify ideas. This process is repeated until the facilitator makes one sheet for each idea being considered. The facilitator posts each idea sheet around the room.

5. The facilitator asks the group to review the advantages and disadvantages of each item. For each item under consideration, participants anonymously rank its advantages and disadvantages on an index card. Completed cards are passed anonymously to the facilitator.

6. The facilitator compiles the ranking and posts the ranking of the advantages and disadvantages of each item on its posted sheet. This will help focus later discussion and help set priorities that must be observed when considering solutions.

7. The facilitator and the group agree on a reasonable time to allow discussion and the facilitator leads a discussion of the advantages and disadvantages of each item and its respective rank. The purpose of the discussion is to clarify and illuminate ideas, and to present the pros and

cons of each item on display, *not* to provide a soap box from which a member "sells" an idea. The facilitator actively manages the discussion to ensure that (a) each participant has equal opportunity to ask questions, seek clarification, provide clarification, or express pros and cons, and (b) that no evaluation is allowed at this point.

8. The facilitator prompts participants to select the ideas or solutions that are most acceptable by ranking them on an index card. Alternatively, the facilitator can assign a specific number of votes and ask the participants to list the ideas and distribute their votes for the ideas to indicate their preference (as in consensus decision making). They can place all votes beside one idea or split their votes among ideas.

9. The facilitator compiles the vote cards and posts the results on the posted item sheets. Ideas receiving no votes can be eliminated. If another round of voting is needed, the facilitator repeats Step 8. For subsequent rounds the facilitator can give participants fewer votes.

10. The facilitator leads a discussion of the selected ideas and may use this opportunity to identify what activities will occur next. For example, if the session's purpose was to identify solutions, the next session may be used to select an approach.

Variations on Nominal Group Technique. The nominal group technique can be used alone as an alternative to consensus decision making. It can be varied to include more group interaction after participants have generated and ranked ideas, or after an idea is selected. McGraw and Seale (1987) found that the combination of the nominal group technique with well-facilitated discussion enhances the creativity and quality of the resulting solutions.

An entire problem-solving process has been built around the nominal group technique and termed the Improved Nominal Group Technique (INGT; Fox & Glaser, 1990). Instead of using nominal groups only for the problem-solving and voting meetings, the INGT suggests a four-phase approach as shown in Table 11.1.

The INGT can help meet any of the following goals:

- Identifying important problems a new system should help solve.
- Defining initial system requirements.
- Reviewing a demonstration scenario.
- Reviewing sample screens or prototype system to identify opportunities for refinement.
- Reviewing requirements documents.

For example, prior to the meeting, the analyst team can present a set of storyboards, sample screens, demonstration or prototype system, docu-

TABLE 11.1
Four-Phase Improved Nominal Group Technique

Phase	Description	Comments
1: Preparation	Goal setting Ideas solicited Numbering and distribution of ideas	Focus is on surfacing and prioritizing problems the group will solve later, and prioritizing problems.
2: Idea Display	Additional ideas collected anonymously Ideas displayed for group	Enables completion and submittal of problems or ideas, if not already done. Enables members to see ideas and propose the elimination, combination, or rewording of an item on display.
3: Discussion	Discussion of ideas conducted Review of all ideas on display	Requires opportunity for discussion of all items. Goal is to encourage participants to speak for or against an item or seek clarification.
4: Voting	Determine nonambiguously the true position of the group on each idea displayed	Determines the appropriate number of most important items to rank. Enables anonymous voting. May require classification of large numbers of items prior to voting.

ments, or other items to group members for review. Each member works individually to document perceived problems with the item. Each of these perceived problems is put on an index card, along with ideas for solving the problem. These cards are brought to a follow-up meeting at which the ideas are displayed and discussed. Finally, the group works anonymously to vote on problems to attack first. Later, they reconvene to consider proposed refinements to the item(s) under consideration.

Using Focus Groups

A focus group is a group of carefully selected participants who meet and focus on specific issues and questions. Goals for focus groups are very specific. Focus groups have been used extensively in marketing and product development to identify and refine requirements for potential products, test design concepts, and evaluate new product features. They are especially useful in the Scenario-based Engineering Process to:

- Refine initial understandings of requirements by asking very specific questions about prototype components' functionality.
- Refine initial design ideas and clarify or extend requirements by getting feedback on "to be" scenarios and prototype screens.

- Validate and refine revised work processes, which will be used to determine required functionality.

Planning the Focus Group. Analysts should be prepared to plan focus groups and be able to communicate those plans to customers, who often do not understand the use of this technique for software projects. Planning includes the following tasks:

- Agreeing on the purpose and goals of the focus group.
- Identifying the desired outcome or output of the focus group and the format in which it will be delivered.
- Determining the kinds of questions to be asked to reach the session goals.
- Developing an agenda and schedule that allots a reasonable amount of time to each issue or question area.
- Identifying appropriate participants.

Figure 11.4 is an example of the introduction and goals statement from a focus group plan.

In addition to the introduction and goals statement, participants feel more comfortable if they know precisely what to expect. Thus, it is useful to provide them with an idea of how long the group will last and the types of activities that will comprise the session. For example, Fig. 11.5 is an

---

Introduction
During the week of April 10–14 a series of three focus groups will be conducted at the hospital as a part of the MSTR project. Dr. Karen McGraw will facilitate these sessions.

A focus group is a group of individuals who meet together to focus on, and provide responses to, specific issues. During each focus group Karen will present ideas for an application in your business function area that would use the database being developed to support the MSTR. She will ask a series of questions and would like each of you to answer them as openly as possible. The two major goals of these focus groups are to:

- Provide a forum for the presentation and refinement of some initial ideas concerning applications that could be a part of the long-term vision in your business function area. These applications would make use of the information residing in the database.
- Enable you to identify data you would require for future applications like those presented to meet your business needs.

FIG. 11.4. Sample introduction and goals statement for a focus group.

Setting:
Central Hospital, Room X14. We will use a roundtable format to conduct the Focus Groups.

Time Requirements:
2 hours

Activities:
1. Introductions and goal setting—statement of session purpose, goals, and expectations.
2. Presentation of initial ideas for process refinement and application support—brief presentation of potential process changes and a sample application in business function area that possibilities.
3. Responding to predefined questions from the facilitators—to identify areas of agreement, areas of disagreement, ideas for refinement, and open issues.
4. Targeting data required to ensure applications meet your business needs—through group activity or individual interviews.

FIG. 11.5. Sample statement providing participants with information on what type of format and activities to expect from the focus group.

example of a statement from a focus group plan that gives participants an idea of exactly what to expect from the session.

Predetermining Questions. Prior to the session the analyst should determine the primary questions to be asked, and should include samples of those questions in the memo sent to participants to help them prepare for the session. Figure 11.6 is an example of the types of questions that would be asked in the focus group. (This example is an excerpt from the focus group plan that analysts prepared for management prior to holding the session.)

Facilitating the Focus Group. At the beginning of the session the analyst posts an agenda and timeline "map" for the session in full view of all participants. As the session begins, the analyst notes approximately how much time will be allotted to each area. The analyst uses a notepad on which questions are divided into the areas of discussion.

During the focus group the analyst uses facilitation skills similar to those described in a subsequent section of this chapter (see *Facilitating the session*). He or she focuses the group on the question to be discussed, using visual aids when reasonable. For example, display a set of sample screens if asking questions to refine screens; display a list of requirements in a particular area if asking about them; display a prototype component if asking questions about its potential uses. Ask a scribe to record comments. The analyst

Sample Questions

During the focus group session, we will ask questions such as those that follow.

- Was the current scenario accurately summarized?
- What changes (to the description of the current scenario) would you suggest?
- We presented a revised process flow description, based on the interviews we conducted in January, in which people in your business function area suggested improvements. What additional suggestions could you make at this time?
- We also presented some areas in which the revised scenario and application *might* provide return on investment. Are these accurate? Is there anything we should add?
- Sometimes a screen image gives you additional ideas about functions the application might provide. Please describe any of these that came to mind.
- Looking at the screen, would each of the field names we used in this image be appropriate? If not, what changes would you suggest?
- We have presented a revised scenario and envisioned a screen to help give you ideas about what a potential application supporting your business function might be like. To operate properly, the application must be able to retrieve the appropriate data from the database and the user.
- This document (conceptual model) represents our current understanding of the data that will be available in the database. Please mark the data you believe you would need and use in an application and process like we described.
- We do not expect that the current set of data is complete. What additional data do you believe would be necessary beyond what you see here?

FIG. 11.6. Provide sample questions to the customer and participants prior to the session.

facilitates the session using questioning and active listening techniques to obtain the depth desired and to clarify inconsistencies and reduce fuzziness.

Perhaps the hardest task in facilitating a focus group is to adhere to the timeline, ensuring that each area gets the focus it requires. We have found it useful to ask the scribe to serve as a timekeeper and let the group know when time for discussion of a particular area has ended. The analyst can then bring questioning for that area to a close and move on.

Analyzing the Output and Documenting Results. The analysis of a focus group entails reviewing any video or audio tapes produced during the session, as well as the notes taken by the scribe. We recommend setting up a worksheet in a word processing package and asking the scribe to enter comments and notes directly into the worksheet during the focus group to reduce transcription and translation time. The worksheet can be as simple as a table that is divided into the different areas for consideration. Each area of the table is then subdivided into the major questions to be asked. Responses are entered in the appropriate area and other considerations or issues that arose can be noted in a "Comments" column beside the question.

The output of a session varies depending on the session goals and the needs of the project; however, it may include the following:

- A formal report, summarizing the findings in each area, responses to each question, and suggestions for using the information.
- An informal document consisting of the worksheet, with an introduction and a conclusion that suggests what refinements or actions should be taken.
- A presentation that summarizes the findings and suggests actions to be taken.

COMPUTER-AIDED GROUP SESSIONS

Networked computers with projection capability, computer-based decision-making support and collaboration tools, and video conferencing technology all can be effective when working with groups.

Networked computers with projection capability, group decision making support, and collaboration tools enable individual members to view, expand, revise, or clarify content that is posted for the group to use in making decisions. Each group participant is provided a workstation that is linked together with other workstations over a local area network, and that provides access to a distributed database. Computer applications for use with groups can be used effectively for brainstorming, consensus decision making, and nominal groups. For example, some organizations have special rooms with networked computers and software such as Chat, a Novell™ groupware utility that allows users to "talk" interactively online with each other, and Visionquest, a group decision-making support application. Another example is TeamRooms, an Internet groupware environment that lets users work together in real-time. TeamRooms provides "shared spaces" in which to meet, and a common place to leave information for others. Using the computer and tools such as these to respond or post information is less threatening for some individuals than speaking to a group and risking the facilitator's rejection or translation of ideas.

This type of arrangement also makes the facilitator's job easier, reducing the need to transcribe the meeting output. Views or data are posted by the members themselves; therefore, the facilitator can manage the group's verbal interactions and time management, rather than translating and posting ideas. Furthermore, actual (i.e., not translated) contributions may be saved to disks, more than one version of a set of ideas may be saved, and saved views may be retrieved for further discussion and revision at subsequent meetings.

Computer-Aided Brainstorming

Computers provide support during brainstorming sessions because they provide the ability to:

- Document participant's ideas quickly.
- Reduce the impact of an overbearing participant.
- Reduce the stress some participants might feel about speaking during a brainstorming session.
- Provide a real-time group view of consolidated ideas.
- Reduce transcription requirements.

Nunamaker, Applegate, and Konsynski (1987) reported success using computers in the decision and planning laboratory at the University of Arizona. Specifically, the research team was interested in how integrated, computer-based systems could help groups solve problems. During interactive brainstorming sessions in the lab, they found that forceful personalities who tend to dominate traditional brainstorming sessions lose their pulpits when the discussions take place in an electronic medium. They instructed participants in the lab to respond simultaneously on their keyboards to questions or issues. Ideas and responses could be automatically compiled and projected to the group without being attributed to any individual.

Even without fancy facilities, analysts can make good use of computers with only an application and a projection device. For example, quality applications exist to support brainstorming, including documenting the session and mind mapping.

Computer Documentation of the Brainstorming Session

Figure 11.7 illustrates results of an initial brainstorming session documented and displayed to the group using Inspiration®.[2] The primary question asked appears at the top of the screen. As each idea is presented, the analyst creates a new symbol and adds the comment. He or she can add text that indicates who provided each comment either during the session or afterwards, as time allows.

Thought Bubbles and Mind Mapping. Thought bubbles are a graphic technique for organizing thoughts, which creates a physical picture of the way we think about something. Mind mapping is a similar process, in which separate thoughts are organized by mapping and clustering them, indicating connections, links, and relationships.

After a process, area, or problem is mapped by a group, the group examines the map while concentrating on the session goal or challenge. Bubbles may be moved around or expanded during this process, resulting in a more complex map. After creatively manipulating the map the group should refocus attention on the problem at hand and seek new solutions. For example, Fig. 11.8 illustrates the mind map produced by a group working

---

[2]Manufactured by Inspiration Software, Inc.

How can the workstation provide better customer service?

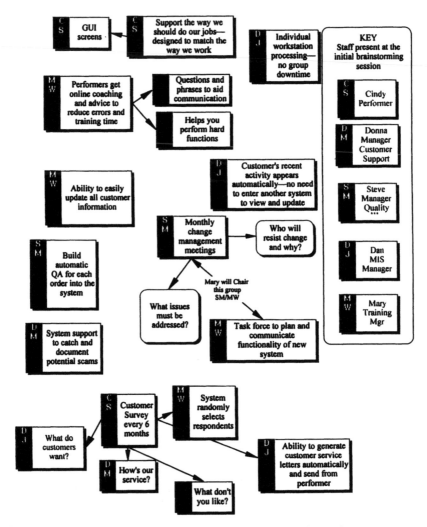

FIG. 11.7. Sample computer-supported brainstorming session.

to identify requirements for a new system to integrate and provide better patient information from point of injury through the aftercare (e.g., after discharge) process. To do so, the group expanded the main points at which information is captured and used. Using the mind map, they brainstormed ideas for helping provide better patient information from each end-bubble (e.g., paramedic tool).

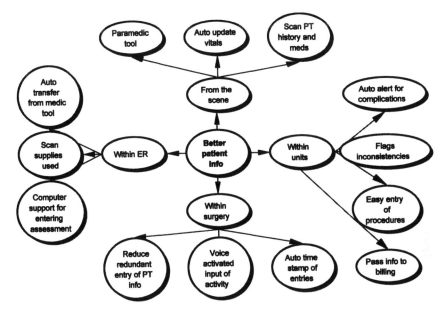

FIG. 11.8. Sample mind map of group brainstorming session.

Computer-Aided Consensus Decision Making

Computer-aided consensus decision making requires special equipment and facilities. Decision support facilities frequently are laid out in a schematic similar to Fig. 11.9. The facilitator's desk contains the equipment necessary to operate the media in the room and interface with the participants' computers. Media in the room usually include devices such as 4' by 6' high-resolution rear projection screens, a video overhead projector, a VCR, networked computers, and decision support applications (Lindwarm & Norman, 1993).[3] The facilitator has access to all participants' machines, and using a video switcher, may display the contents of any screen on the facilitator's screen. A group decision support application offers a collaborative environment for sharing ideas anonymously. Various other tools may be available, including brainwriting applications and bulletin board applications for online discussion.

The decision and planning laboratory at the University of Arizona is structured in a similar manner and enables participants to review others' responses, append their ideas, and work toward consensus. Software consolidates comments and organizes them for display on a large screen. Nunamaker, Applegate, and Konsynski (1987) described the positive in-

---

[3]One example of this type of system is the AT&T Teaching Theatre at the University of Maryland, which supports 20 computers on a network linked to a Novell™ server.

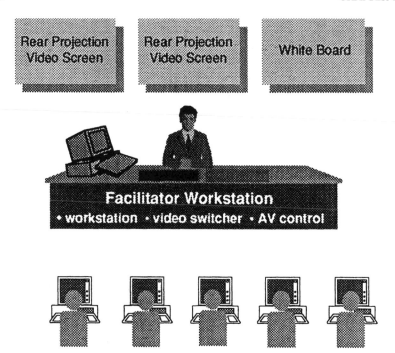

FIG. 11.9. Sample schematic for computer-aided decision-making support
facilities.

fluence of computer facilitation of group decision making by stating that
"Computers force you to concentrate on the problem and allow you to
express yourself. The system lets you focus on ideas rather than on the
people who came up with them" (p. 5).

Another example of the use of computer-aided decision making is Xerox
Parc's Colab. Colab uses dynamic, interactive media and computers to
increase the effectiveness of group sessions (Stefik, Foster, Bobrow, Kahn,
Lanning, & Suchman, 1987). Available Colab tools provide participants
with a coordinated interface that allows members to interact with shared
objects on the screen and public, interactive windows accessible to other
group members. Colab also provides private windows with limited access,
and supports simultaneous action, allowing group members to work in
parallel and act on shared objects.

Using Computer or Video Conferencing

Computer or video conferencing provides a fertile environment for
requirements elicitation, analysis, and review sessions. Figure 11.10 illustrates the concept, in which individual or multiple participants at two or
more sites communicate using video conferencing technology.

FIG. 11.10. Video conferencing as a tool to support group sessions.

McGraw (1991) successfully used video conferencing on a NASA project that required elicitation sessions with expert quality engineers from numerous NASA sites. Due to a tight travel budget and the inability to pull so many experts from their jobs for travel to a central location, video conferencing was used to support the sessions. Selected experts (3–4 people) at each of six NASA sites were asked to participate and notified of exact session goals, date, and time. They received a packet of information for review prior to the session. The session was facilitated by McGraw and an analyst team from Johnson Space Flight Center. The video conferencing system allowed the analyst team to scan from site to site at predetermined intervals (e.g., every 5 minutes), enabling them to see all participant groups equally. If a participant responded to a question, was asked for specific information, or asked a question, the view could be switched to that site, enabling the analysts to see the speaker. When any participants wanted to share data or sketches, they were able to project it to the other participants using a separate device. Positive outcomes of this session included:

- Better cohesiveness among a diverse, geographically dispersed group.
- Inability for any one person to dominate the session.
- Facilitation enabled by technology—the analyst team was the primary site and could control whether the projection of other sites was random, on 5-minute intervals, or performed in response to a question or comment.
- Data, graphics, and ideas could be shared visually as well as verbally. The ability to project this information to the team was critical to the

success of the meeting because it provided a means to make use of more channels of communication and enabled shared communication (i.e., the posting of ideas to a list everyone could see).

- Travel costs and experts' time were managed more effectively, saving the project money and making the best use of the time the experts had available.

MANAGING GROUP SESSIONS

Many people do not feel comfortable managing sessions. However, careful attention to the following can greatly enhance the ease with which group sessions are conducted:

- Careful selection of participants.
- Facilitating and managing group session.
- Debriefing participants.

Selecting Participants

Carefully select participants for group sessions. Table 11.2 identifies interaction skills analysts should look for in effective candidates for group sessions.

When selecting participants, analysts should also consider the impact of group size on the effectiveness of group sessions. Although it is difficult to pinpoint an ideal size, it is known that groups with fewer than five members generate more discussion and higher levels of participation. Groups larger than seven members tend to experience more member inhibitions, possible domination by aggressive members, and a greater tendency to split into subgroups (Schermerhorn, Hunt, & Osborn, 1991). Thus, as group size increases, session effectiveness decreases, as does the ease of facilitation.

TABLE 11.2
Characteristics of Desirable Group Participants

Tactful	Enthusiastic
Sense of humor	Cooperative
Friendly, open	Minimizes differences
Interacts easily with others	Can identify with the session's goals
Works to help group succeed	Considers rewards of membership in the group

Adapted from Sattler & Miller (1968), *Discussion and Conference*, pp. 221–223, and McGraw & Harbison-Briggs (1989).

Facilitating the Session

Facilitating any group session requires multiple skills and mental processing on the part of the analyst. If a group session is to be effective, the analyst must be able to function as an effective group leader and facilitator. Primary skills relevant to group sessions include the following:

- Setting the stage and expectations by presenting the session goals in a clear, concise manner.
- Keeping the purpose or goals of the session in view.
- Refocusing participants as needed to keep the session on track.
- Facilitating discussions among multiple participants to ensure that each has the ability to express ideas and be heard.
- Mediating negative or controversial discussions to ensure that each participant retains respect for the others and maintains a positive attitude.
- Developing and building a team environment.
- Dealing effectively with the occasional problem participant.

In addition, analysts facilitating group sessions must have effective communication skills, including the following (see chapter 4):

- Good questioning techniques.
- Effective use of active listening, including rephrasing and clarifying ideas.
- Ability to hear, and quickly sort through ideas, and tie seemingly disparate comments together.
- Monitoring the nonverbal behavior of participants.

Debriefing Participants

Debriefing is especially important when working with group sessions, and includes the following activities:

- Clarifying and refining information presented during the session.
- Defusing potential problems with interpersonal relationships.
- Validating the information presented.
- Giving participants the opportunity to voice concerns.
- Getting buy-in from each participant.

Analysts should perform two types of debriefing—a formal debriefing with the entire group at the conclusion of the session, and an informal one with each individual member of the group. The formal debriefing occurs during the closing of the session and summarizes (a) session goals, (b) key areas discussed, (c) ideas presented, rejected, refined, and/or accepted, (d) key unresolved issues, and (e) opportunity for performers to voice their support for the session output.

The informal debriefing involves a brief conversation with each member to ensure that they were able to freely express agreement or disagreement with the session output without fear of embarrassment or compromise. Confidentiality is ensured, valuable information may be acquired, and the performer's willingness to participate in future sessions is enhanced.

TYPICAL PROBLEMS OF GROUP SESSIONS

Factors that can negatively affect the productivity of a group include member equality, rank and status issues, confidentiality, consensus versus diversity, and access.

Rank and Status Issues

A person's status indicates his or her relative rank, worth, standing, or prestige in a group (Schermerhorn, Hunt, & Osborn, 1991). Status can be based on age, work seniority, accomplishments, or rank in the organization. If members of the group have equal or similar status, the group will be easier to facilitate and manage. On the other hand, if a participant is outranked or even managed by another in the group, problems may occur both in managing the group and eliciting valid content. Each time the lower-status participant responds, he or she is likely to glance toward the more powerful member for approval, then amend the original statement based on the perceived reaction. McGraw and Seale (1987) termed this phenomenon *upward ripple paranoia.*

To combat the effect of unequal status, analysts must carefully select group participants to reduce the opportunity for power struggles and fear of reprisal. Furthermore, good communication, the clear presentation of the session goals, and the effective use of group facilitation skills can help alleviate the problem. The atmosphere in the group session must make each participant feel comfortable contributing and evaluating ideas.

Confidentiality

Another issue that must be addressed is confidentiality. Participants should not have to worry that their contributions will be shared and evaluated by other participants outside of the session, unless the idea stands alone and is not attributed directly to him or her. Facilitators help ensure confidentiality by stating at the beginning of the session exactly how the information will be used and, if it will be shared with others, how confidentiality will be handled.

Diversity Versus Consensus

It is no surprise that performers in organizations that use teamwork and empower workers are easier to work with when group consensus is required. Obviously, it is easy to facilitate groups in which there is general agreement or few political or organizational issues that impede the ability of the group to work together and reach agreement. Even when multiple performers provide multiple opinions, diverse ways of looking at problems and solutions, and some conflicting information, people who are used to working in teams will be easier to manage in group situations.

Group sessions in organizations with infighting, interdepartmental distrust, group disparity, and recent layoffs require much stronger facilitation and leadership skills on the part of the analyst. Diverse individuals or groups may also be difficult to move toward consensus. In these sessions the facilitator must determine how hard to push the group toward consensus, and carefully facilitate session goals.

Different participants may offer diverse responses and should not be forced toward consensus. As Reboh (1983) noted, even if two equal-status performers arrive at the same conclusion or solution, they may not have used the same reasoning process. Analysts must determine whether the goal is consensus and group buy-in, or whether to identify the points of disagreement and diversity as targets for later investigation and manipulation.

One technique used in knowledge-based systems to handle conflict resolution that may be useful is the meta rule concept (McGraw & Harbison-Briggs, 1989). Meta rules enable conflict resolution using a higher-order policy heuristic. Meta rules are usually hierarchical in nature—the analyst must elicit or find the higher-order rule or guideline that would operate on *all* possible conflicts, and must be able to determine when more specific guidelines apply.

Other approaches to allowing conflict in a system include the use of fuzzy sets (Klir & Folger, 1988) that ascribe categorical values of belief (e.g., good, greater than, etc.) to a derived guideline or suggestion, and uncertainty methods (Kanal & Lemmer, 1986) that establish calculations or certainty values for a guideline, suggestion, or rule.

Access and Participation

Finally, obtaining access to even a single expert performer can be difficult—this problem is compounded when multiple performers are requested for a group session. The customer may balk at pulling key performers from their positions for a 2-hour group session. Analysts must be prepared to convince customers of the usefulness and criticality of the group session to gain access to required performers. Once access is granted, the date and time for the session can be determined. It has been our experience that group participants are more likely to attend the session and arrive at the location on time because they know that others also are committing their time (Hunter, 1985).

Sometimes the customer agrees to make the performers available, but fails to relieve them of any responsibilities such as modifying their daily performance goal (McGraw, 1994b). Consequently, the performers may be reluctant to participate. Analysts can help reduce the potential for performer reluctance by suggesting to the customer that the daily performance goal be modified, or that performers account for their time using an indirect labor code, if that is an issue.

Chapter 12

Evaluating and Refining Requirements

The completeness and quality of a software application depends on many factors. Software developers may consider an application to be of high quality if it has characteristics such as correctness, flexibility, testability, reliability, efficiency, and reusability (Pressman, 1992). Although developers are more likely to focus on technical quality (i.e., the "-ilities") and cost, Jenkins (1995) noted that customers or users determine software quality based on an assessment of the degree to which the software product affects their work or life. The following operational characteristics are considered critical by customers and users:

- The software product possesses the desired combination of attributes and features (IEEE, 1983).
- The software product fulfills a stated purpose (Reifer, 1985).
- A customer or user perceives that the software product meets his or her composite expectations (IEEE, 1983).

Obviously, both perspectives (i.e., technical and operational) must be integrated if we are to determine a product's overall quality and increase its percentage of use in the "real" world.

Many software quality issues, especially those viewed from a customer's perspective, are impacted by the requirements phase of a project. During this phase, the problem domain is analyzed for the purpose of understanding user requirements well enough to produce software requirements that are reasonably correct, complete, and consistent. This phase is critical because of the origins of software defects—requirements, design, coding, documentation, and bad fixes—*requirements* errors that are among the most costly and difficult to remove (Jones, 1995). Common problems that lead to defective requirements include:

- Developer assumptions that the requirements are "obvious."
- Failure to do enough requirements elicitation and analysis (Lawrence, Johnson, & Embley, 1995).
- Failure to include appropriate personnel (i.e., users, customers, shareholders, operations personnel, engineers) in requirements elicitation and analysis activities.

Additionally, it may often be the case that the requirements specify *what* a system should *do*, but do not detail *how* the product will be *verified.*

SEP was developed on the premise that a high degree of customer/user involvement, use of customer scenarios to bound and reveal requirements, and iterative refinement of requirements (through prototypes and demonstrations) will result in more complete requirements and thus, better software applications (see chapter 1). Proofs-of-concept and/or prototype systems, and demonstration systems help developers state requirements in quantifiable terms and specify how the product should be verified. Furthermore, evaluation of elicited requirements is ongoing and constant throughout SEP (see chapters 1 and 3).

This chapter discusses evaluation within SEP. First, the role of ongoing evaluation is presented to help the reader understand all of the points in SEP at which evaluation occurs. Next, different approaches to evaluating the total set of preliminary requirements are highlighted, because the approach selected impacts the types of evaluation completed. Finally, techniques for evaluation are presented. As the reader discovers, evaluation mechanisms should be woven throughout the use of each elicitation or analysis technique presented in this text.

Specifically, this chapter is designed to help the analyst achieve these goals:

- Understand how SEP builds in evaluation throughout the elicitation and engineering process.
- Recognize the approach most appropriate for a specific project.
- Understand the variety of evaluation mechanisms available.
- Be able to use evaluation techniques with appropriate participants to refine requirements.

THE ROLE OF ONGOING EVALUATION IN SEP

The process of evaluation is a major means of assuring software quality and promoting confidence in the system. Evaluation is most often achieved through verification and validation (V&V). *Verification* is the process of assuring that the product is *built right*. *Validation* is the process of assuring that the *right product* was built. In SEP, both verification and validation are incremental in nature. Evaluation is applied at each step of the process to incrementally refine the requirements, behavior, and characteristics of the components. Figure 12.1 conveys the concept of V&V in SEP.

Developers work with users, shareholders, and project champions to identify scenarios that represent an important part of the problem space. During knowledge acquisition or requirements elicitation sessions, analysts use the appropriate technique(s), such as observation, task analysis, etc. After each session the analyst reviews the information acquired and writes a session summary report. The report is reviewed for accuracy and completeness by the user or other person who participated in the session. The information acquired is analyzed and used to develop preliminary requirements. Types of requirements that should be identified through this process include those shown in Table 12.1.

Next, a prototype or demonstration system is designed from these requirements and envisioned scenarios. The prototype or demonstration system is implemented based on the envisioned scenarios, preliminary requirements, and screen designs. These activities not only help refine the requirements, but also help developers specify requirements for interfaces, resources, portability, reusability, documentation, maintainability, and safety. Throughout prototype or demonstration system design and implementation, users, shareholders, champions, and developers are involved in evaluative reviews and activities.

APPROACHES TO ITERATIVE REQUIREMENTS EVALUATION AND REFINEMENT

Thorough requirements analysis and specification requires iterative, ongoing evaluation and refinement of requirements. Consequently, evaluating the completeness and accuracy of requirements must occur at each step in the process. In SEP either of two approaches may be employed to integrate, evaluate and refine requirements: (a) proof-of-concept and/or prototype system, or (b) demonstration system. Incremental verification and validation should occur regardless of the iterative approach selected.

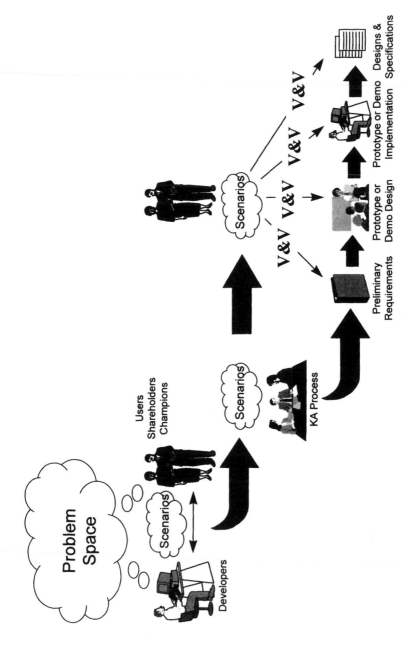

FIG. 12.1. Evaluation—V&V—within SEP.

TABLE 12.1
Examples of Preliminary Requirements

Requirement Type	Example
Functional	*The software must enable creation of new customer accounts.*
Operational	*To create a new account, the user interacts with the Customer Account screen. Figure 1.1 presents the layout of the screen. Table A describes the actions the user can take from this screen.*
Performance	*The system must respond to user input (i.e., button press, command selection) within 2 seconds.*
Acceptance	*The system will be validated using test cases built from scenarios identified during knowledge acquisition and approved by the customer.*
Security	*The system will be protected by a two-level password system.*
Reliability	*Mean time interval between failures will be XX.*
Verification	*The software will be verified using simulated tests, followed by live tests with real inputs.*

The development of a proof-of-concept and/or prototype system is the more common approach, and is appropriate for projects that are fairly well understood and are only moderately complex. The proof-of-concept or prototype system is developed based on scenarios and other elicited information. During its development it is evaluated and refined until it reflects requirements that are believed to be as complete and accurate as possible.

The development of a demonstration system is more costly and cumbersome; consequently, this approach is reserved for very complex, large, or poorly understood system efforts. Using this approach the development team first focuses on a *subdomain*, or part of the total domain. Working within this subdomain, they develop a demonstration plan, script, and system that can be exercised and evaluated in mock situations, or in situations that appear to be like the normal environment. After the subdomain is well understood, developers will repeat the process for another part of the domain.

The creation of either a prototype *or* demonstration system requires that the verified[1] products from requirements elicitation sessions (i.e., task/subtask descriptions and constraints, responsibilities, etc.) be analyzed. This analysis, and the subsequent changes in the way the information is represented, results in a set of reference requirements and domain models as shown in Fig. 12.2. In addition, developers begin to understand the constraints and responsibilities that the system architecture must address. At this stage, the requirements are generalized abstractions of user and expert definitions of what the system can do, and how well it should do it. The

---

[1]Products are verified by the person/expert participating in the session, and possibly, by a panel of experts or by project champions, to ensure accuracy.

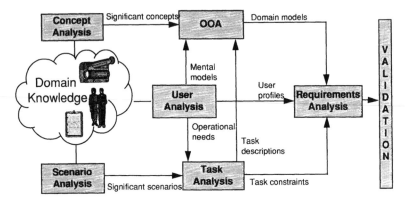

FIG. 12.2. Validating specifications for use in prototypes and demos.

workable domain model is a description of the domain represented as objects with attributes, services, and messages (Kelly & Harbison, 1995). As the prototype or demo system is further defined, feedback is elicited from users, project champions, and shareholders. Thus, the domain models, reference requirements, and reference architecture are continually validated. The prototype or demonstration system development process provides further validation of these requirements and models, and identifies needed refinements. Each of these two approaches is presented in further detail.

Proof-of-Concept and Functional Prototype Development and Evaluation

A proof-of concept or prototype is an incomplete, initial version or "pilot model" of the intended system (Asur & Hufnagel, 1993). It represents partial functionality and usually does not meet nonfunctional requirements, such as performance. However, it *does* help reveal and clarify operational and functional requirements, which enables developers to further investigate other requirements, such as performance and resource requirements.

There are many benefits to the development of prototypes, such as:

1. Reduced deadline pressure.
2. Prototyped products tend to result in a better user interface (Boehm, Gray, & Seewaldt, 1984).
3. Prototypes result in reduced cost.
4. Prototypes help developers find and specify user requirements, which improve the quality and completeness of requirements.
5. Prototypes help bridge the requirements–specifications communication gap.

6. Prototypes enable project teams to study feasibility and identify and resolve potential problems.

If a prototype system approach is used, the project team will be continually evaluating their knowledge acquisition, requirements analysis, and initial design products. Ongoing evaluation with users, shareholders, and project champions occurs throughout the iterative prototype development cycle. For example, Fig. 12.3 illustrates evaluation within a prototyping approach.

Note that after the prototype is implemented and evaluated, iteration is based on the type(s) of problems identified during the evaluation. For example, if problems were based on missing or incorrect requirements, additional sessions observing users or refining a task analysis may be necessary. If problems are more related to usability issues, iteration may involve revising screen designs or content on a screen.

Eliciting and Analyzing User Requirements

First, preliminary knowledge acquisition occurs, in the form of observations, domain familiarization, discussion of scenarios, and initial structured interviews (see previous chapters in Part II). These sessions may

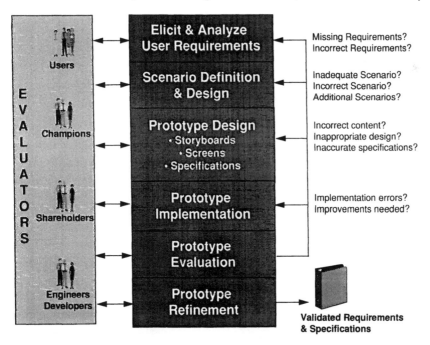

FIG. 12.3. Evaluation within the proof-of-concept and/or prototyping approach.

include surveying the potential user population to define user profiles or characteristics, and will most certainly include sessions that help reveal functional or operational requirements.

After the analysts have a grasp of process and problems, they may convene a small group to refine their findings and initial ideas. Typically this is followed by a meeting with project champions, at which the goals, focus, and functionality of the prototype are evaluated and further defined. This includes defining the scenario(s) that will be used to mock up screens and functions.

Defining and Designing Scenarios

Next, the scenario(s) for the prototype are mapped out, and the screens necessary to follow the scenario through to completion are identified. This is followed by a review by project champions and carefully selected users. As these individuals review the scenarios and the screens that will be developed to support the scenario, they begin to understand better what the system might be able to do, and are frequently able to provide additional ideas such as other functions desired, additional screens that should be developed, clarification about a domain area or function, and so on. The goal of this review is to refine the developer's understanding of requirements for the system at a stage during which additions or changes to the requirements are not prohibitively expensive. Throughout this process the development team learns more about the requirements for the application. One output of this step is a preliminary software requirements document.

Designing Prototype Storyboards, Screens, and Specifications

Prototype design occurs next. This involves use of a prototype description language, or charts, diagrams, and screen sketches. At this stage the prototype begins to become "real" to the users and project champions. After drafts of screens are complete, they are reviewed and evaluated with the project champions and selected users to ensure usability and completeness of the prototype. As they see screen content and discuss desired functionality, additional requirements and ideas for modifying preliminary requirements are revealed and detailed. For example, users and project champions may identify missing fields, data, or selections on a prototype screen; may identify a missing function; or may simply provide corrections and clarifications.

Implementing the Prototype

Next, the prototype is implemented as designed. The emphasis here is on rapid implementation with low development costs. After the prototype has been implemented, it is evaluated by users, shareholders, project champions, engineers, and developers. Each of these individuals provides feedback that is used to extend, enhance, or identify requirements. The output of this process is incorporated into the requirements specification document (Asur & Hufnagel, 1993).

Refining and Evaluating the Prototype

Refinement of the prototype continues as it is evaluated to determine its *actual* behavior, as compared to its expected or desired behavior. As changes are made to the prototype, the requirements specification document is updated. The process of iterative refinement continues until the prototype captures the critical aspects of the envisioned system. The refined requirements and specifications, which were produced as a result of this process, are used to design and implement the production application. The other product of this process—the prototype—can be used to share the "vision," and to support organizational change.

Demonstration System Development and Evaluation

Prior to full system development of a very complex system, a project team may choose to develop a demonstration system. The main goals of a demonstration system are to facilitate an understanding of the domain (including user requirements) and performance, interface, resource, and other critical requirements issues (Harbison, Burnell, Kelly, & Silva, 1995). Demonstration system development and evaluation is illustrated in Fig. 12.4 and is described in more detail in the subsections that follow.

Selecting the First Subdomain

The primary reason to use a demonstration approach is that the domain is very complex. Developers can plan for a series of demonstration systems, each of which focuses on a portion of the domain and builds on previous demonstration efforts. The first task the team must undertake is to decompose the domain into areas of smaller scope and select one of the subdomains for the development of the first demonstration system. For example, in the trauma care domain, the scope may be decomposed into military, urban/civilian, and rural subdomains, each of which could be further decomposed. For the purposes of a demonstration, the team would select one of these subdomains for the first demonstration. Other subdomains could be addressed in subsequent phases of the project. Doing this fosters incremental domain understanding and allows progress to be made in the initial phases of development, when the domain is too overwhelming to tackle in one piece.

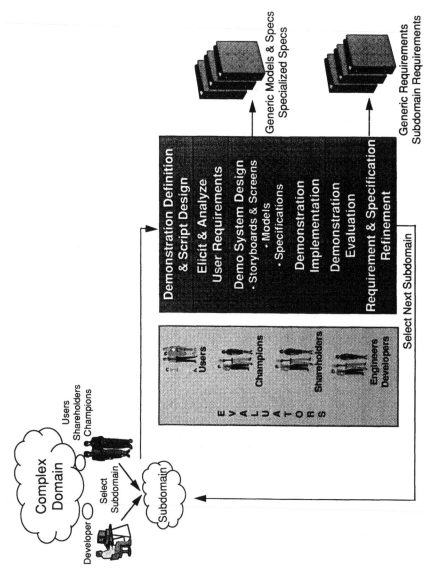

FIG. 12.4. Demonstration system development and evaluation is effective when the domain is very large and very complex.

Defining the Demonstration

After the project team selects a subdomain, they meet to define the goals of the demonstration and identify the factors that will impact its perceived success. For example, Fig. 12.5 is an example of critical success factors for a civilian trauma care demonstration.

Working from the goals and critical success factors that have been defined, the project team ranks the goals and factors to determine which are most important to meet, which are least important, and so on.

Developing a Demonstration Plan

Next, the project team develops a demonstration scenario plan, structure, or outline. A demonstration scenario may actually consist of a number of separate "scenes" to ensure that the primary goals have been met. Initially, the plan, structure, or outline need not be more than descriptions of each scenario in the demonstration and its conditions, factors, and constraints.

Capturing Scenarios and Requirements

Using this structure as a foundation, the project team begins using the techniques described previously in this text to capture scenarios, analyze tasks, and so on, to enable them to understand how the demonstration system must operate. The purpose of this activity is to define an appropriate specification of the demonstration system to be constructed. This specification, along with scenarios for the demonstration, are the starting points for the synthesis of the demonstration application's architecture.

✔ Demonstrate more effective flow of information from the field providers to the hospital.

✔ Demonstrate a reduction in overtriage of patients, to ensure resources are applied most effectively.

✔ Demonstrate the ability to capture information in the field via voice and pen input in a variety of environments (e.g., noise, light, etc.).

✔ Demonstrate the ability to handle normal incidents (1–2 patients) and integrate with other systems to handle mass casualty situations (e.g., multijurisdictions, etc.).

✔ Provide support for hazardous materials emergencies.

✔ Demonstrate protocol support for infrequent interventions and treatments.

✔ Demonstrate the ability to support both medical and trauma emergencies.

✔ Demonstrate the ability to display and alert the paramedic to degrading patient conditions (i.e., Airway, Breathing, Circulation).

✔ Demonstrate easier, faster, more accurate completion of patient-related reports.

FIG. 12.5. Sample critical success factors for civilian trauma care.

As scenarios are elicited, tasks analyzed, performances observed, and sample screens developed, the project team begins to better understand the domain (i.e., its vocabulary, concepts, processes, and goals). The requirements elicited and screens designed for this demonstration system will reflect an incomplete picture of the entire domain, however, and will be somewhat simplified in nature.

Developing a Demonstration Script

At this point the team can begin the development of a demonstration script. At first, it is likely to be little more than a structure or framework for the eventual script. Initial information in the script should include the name and identifying number of each major scene or event, characteristics and features of each scene or event (e.g., roles involved, number of participants/players, system involved, setting), and the major storyline or scenario that is depicted in the scene, as shown in Fig. 12.6. Later, the script will be fleshed out to include a narrative that describes everything that happens in each event, including activities or tasks completed by each role, system activities and responses, windows and messages that appear on the screen, and so on.

Evaluating the Demonstration Script

After a preliminary script has been developed it should undergo evaluation by a carefully selected group of shareholders, experts in the domain,

Event 1: Customer with billing problem	
Narrative description:	Customer calls in disputing 3 charges on MasterCard® and demands immediate resolution
Setting:	Early in the day; unspecified call center
Goals/critical success factors:	Handle call on 1–2 screens Handle call without transferring call Obtain system assistance in handling irate caller Maintain acceptable call handle time
System functions used:	Customer Screen History Window "Tips for handling irate callers" help file
Roles involved:	Service agent Caller/customer
Characteristics of each role:	Service agent: New on the floor; just completed training Caller/customer: High value customer; irate

FIG. 12.6. Sample script skeleton.

potential users, project champions, engineers, and developers. We have found it useful to include people who have participated in knowledge acquisition sessions in this script review because they already have a sense for the demonstration system's goals, purpose, and general content. The person responsible for managing the development of the demonstration script sends it to the reviewers, and asks each to read it and complete a questionnaire about it prior to the review meeting. (The questionnaire enables the developer of the demonstration script to receive subjective information about the quality and completeness of the script, and its ability to meet stated goals and objectives.) Later, a review meeting is held and the demonstration script developer requests the completed questionnaires. After reviewing questionnaire responses and comments, he or she facilitates a walkthrough of the script. During the walkthrough, the group is asked to comment on inaccuracies, offer constructive criticism, identify opportunities for refinement, and provide necessary clarification and information. All suggestions are captured for later consideration. In the event of conflicting suggestions, project champions are responsible for suggesting the resolution to the demonstration script developer.

Implementing the Demonstration

The development of the demonstration system involves a series of activities, including:

- Validating knowledge acquisition session information.
- Defining languages for use.
- Analyzing performance goals and system composition rules.
- Selecting the components the demonstration system will require.
- Developing a demonstration system design.
- Constructing components.
- Specifying language and standards.

Prior to conducting the demonstration, the demonstration system undergoes numerous reviews and is evaluated again by the development team and a panel comprised of shareholders, users, and project champions. Each time a review is conducted, information that can be used to improve and refine requirements is captured, documented, and to the extent feasible, used to change the demonstration script or system.

Evaluating the Demonstration System

The demonstration system and the entire development process are then evaluated by users, shareholders, developers, and others who attend the demonstration. Screens and system functionality clarifies or reveals requirements errors and omissions. This usually stimulates participants to provide additional information or ideas. In addition, the demonstration can reveal misunderstandings of the domain, or errors in the domain model, components, or scenarios. The evaluation of the demonstration should address subjective issues, such as how well the participants liked the demonstration and individual features of the system, *and* objective issues, such as whether they believe the system met specific requirements.

After the demonstration system has been evaluated, the project team may select another subdomain for which to develop a demonstration system. As subsequent demonstration systems are developed, domain models, requirements, specifications, and components common to the previous demonstrations are identified and refined. In addition, domain models, requirements, specifications, and components that are specific to a particular subdomain are identified. The team uses the task specifications, components, experience, questionnaires, and other products output from the demonstration development and evaluation process to facilitate the development and refinement of requirements for the entire domain.

MECHANISMS FOR EVALUATION

Regardless of the approach selected, there are many ways to verify that the right information has been acquired and represented appropriately. Table 12.2 presents a sampling of evaluative techniques appropriate for each step in the process, and suggests possible evaluators.

As shown in Table 12.2, reviews, walkthroughs, and other mechanisms are used throughout the evaluation process. The sections that follow describe important evaluative activities, and the techniques used in each, in more detail. These include:

- Reviewing the plan to identify requirements elicitation sessions.
- Evaluating session output and products to verify the correctness of information elicited during a session.
- Conducting group reviews with project champions, shareholders, and users.
- Conducting operational design walkthroughs.
- Conducting usability evaluations.
- Conducting paper-and-pencil evaluation.
- Using questionnaires and surveys.
- Conducting field studies.

TABLE 12.2
Evaluative Mechanisms and Evaluators Appropriate for Each Step

SEP Step/Process	Evaluative Mechanism	Possible Evaluator
Create and apply a knowledge acquisition or requirements elicitation plan; identify required sessions	Comparison of a desired session to the knowledge acquisition plan	• Analyst, Requirements Manager
Define or identify scenarios	Scenario reviews	• Users, shareholders, champions
	Preliminary feasibility	• Developers
Requirements elicitation activities (observation, task analysis, etc.)	Review of session notes/tapes	• Analyst, Requirements Manager
	Review of completed Session Report to verify correctness of information	• Subject matter expert • Analyst • Expert panel
Create envisioned scenarios	Scenario reviews	• Users, shareholders, champions • Developers
Create storyboards/screen drafts	Focus group review	• Users, shareholders, champions
	Walkthrough	• Users, shareholders, champions
	Paper-and-pencil evaluation	• Developers • Users
Develop Prototype or Demo system screens	Walkthrough	• Users, shareholders, champions
	Focus group review	• Users, shareholders, champions
	Survey	• Users
	Expert evaluation/review	• Users
	Usability review	• Users
Develop Prototype or Demo system	Field study	• Project champion, users
	Usability review	• Project champion, users

Review the Plan

Before conducting requirements elicitation sessions, customers, shareholders, and users should review the knowledge acquisition or elicitation plan (see chapter 3). A team comprised of these individuals can help identify oversights, avoid the omissions of important information early in the process, and help determine the optimum sequence for the elicitation and analysis of specific information. Furthermore, they can help identify the person(s) who can best provide high quality information about specific, target content areas. As analysts begin to conduct sessions, they should confirm that the desired session helps meet the goals outlined in the knowledge acquisition plan.

Evaluate Session Output and Products

As Fig. 12.7 illustrates, the knowledge acquisition plan, requirement elicitation technique used, session goals and coverage, session management (i.e., keeping it on track), and session participants (analyst, subject matter expert, user, etc.) all impact the quality of information and requirements identified. Theoretically, one should be able to trace requirements back to a session report or set of sessions, which enhances our confidence that we are indeed "building the right system." Evaluation of session-related information involves adhering to the knowledge acquisition plan and verifying the correctness of session information.

Verify the Correctness of Information Elicited During a Session

Several types of useful reviews are possible at the conclusion of a requirements elicitation or knowledge acquisition session. First, the analyst reviews notes taken during the session and listens to or views tapes made

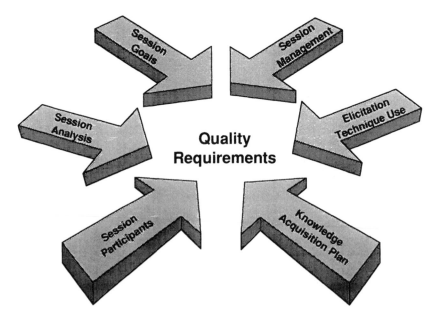

FIG. 12.7. The quality of requirements is impacted by a number of factors related to requirements elicitation and analysis.

of the session. The information elicited is analyzed and organized for presentation in a session report. During this process the analyst is conducting an initial check of the information by confirming correctness (as he or she heard and interpreted it) and clarifying questionable areas.

After the analyst feels confident in the information and the way it is organized and presented in the session report, he or she should ask the subject matter expert (or other participant) to review it. As noted in chapter 3, reviewers are encouraged to edit the report and offer suggestions for refinement and corrections. This aids in certifying and verifying that the important information has been captured, interpreted, and presented correctly. Only when the information in a report has been deemed "accurate" should it be posted on the server, entered into a database, or used to extract requirements. Figure 12.8 illustrates this complete process, which ensures that requirements are based on accurate information and are traceable.

Conduct Group Reviews

Focus groups typically are used in the early stages of requirements elicitation, while other groups may be convened to review developing scenarios or screens (see chapter 11). Group participants may be called on to evaluate scenarios, draft screen sketches, storyboards, or actual prototypes. The objective is to identify the acceptability and completeness of a concept or item and explore ways to enhance acceptability. Examples of group reviews include project champion, shareholder, or expert evaluation, user evaluations, and actual demonstration events.

Project Champion, Shareholder, or Subject Matter Expert Evaluation

Evaluation by project champions, shareholders, and subject matter experts is usually conducted against a specific set of requirements, goals, or critical success factors for the project. The viewpoint should be twofold, that of (a) the specific target population who will use the product, and (b) overall impact to the organization. Reviews by champions, shareholders, and experts may be conducted individually or in focus group formats. This usually entails presenting the product (prototype, demo system screens, etc.) and asking questions related to specific portions of the product. For example, the facilitator might demonstrate a set of screens relating to a specific function, then ask participants specific questions. After responses have been elicited, the facilitator could ask for additional comments before moving on to the next review area. The purpose should be to ensure that the developing product will meet the needs of the target population and do so in a manner than has positive organizational impact. As noted in

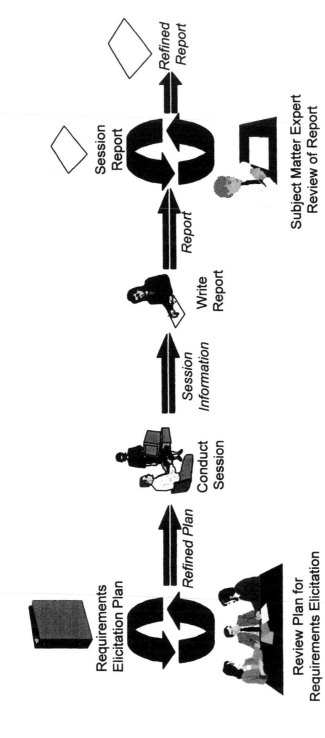

FIG. 12.8. Evaluating session information and enabling traceability.

chapter 11, however, there are some considerations for using a panel or group comprised of a number of champions or shareholders. These individuals are typically "experts" in their own area of the domain. Consequently, this group is prone to problems such as member equality, conflicting information, and biases, and requires a skilled facilitator to ensure positive outcomes.

User Evaluation

The ultimate criterion for success is whether a system built according to the prototype could be used effectively by the target population. Involving representative end users in the review process helps ensure that the system being specified and designed actually will meet the needs of the people for whom it is being developed. An evaluation by end users may include:

- Reviewing scenarios to ensure they cover the domain appropriately.
- Reviewing draft screen sketches and developing prototypes or demonstration systems to determine whether all of the desired functionality has been specified.
- Ranking requirements to enable developers to explore trade-offs and phased builds.
- Reviewing the usability of prototype screens or demonstration systems.

Conduct Operational Design Walkthroughs

Operational design walkthroughs are commonly done by developers, designers, or human factors engineers working with representatives from the target population. The purpose of an operational design walkthrough is to envision the user's route through a proof-of-concept or early draft screens (prototype or demonstration system) and identify potential problems or concerns (Rubin, 1994). Participants involved in this review might include project champions, developers, designers, and one or two sample users.

To ensure that the group does not become bogged down in this process, it is important that operational design walkthroughs are carefully planned. Planning should include confining each walkthrough to one function or operational area at a time, providing participants ahead of time with a copy of screens or flowcharts to be presented, and limiting each session to no more than two hours.

Usability Evaluations

Usability evaluations capture data relating to ease of use of a prototype or demonstration system, including the following (Rubin, 1994):

- Effectiveness—speed, performance, error rate.
- Usefulness—meets operational goals.
- Learnability—ability to operate system.
- User attitudes—how users feel about the system and its interface.

During usability evaluation, the following areas may be addressed— screen navigation and information flow, screen layout and object selection, menu and command use, display issues, and data entry (McGraw, 1992). Usability evaluations start with reviews of screen sketches to enable developers to capture important usability information as early as possible. However, most of the information provided by this technique occurs as the prototype or demonstration system is being implemented. Key questions to consider during usability evaluations include those shown in Table 12.3.

Data collected during a usability evaluation can be categorized in two primary ways: subjective data reflecting users' preferences, and objective data reflecting the performance of participants. Preference data will vary according to the product, but will usually include information such as which version of a screen, window, or icon they preferred, identification of things they do not like or think are easy to use, and comments or suggestions for improvement. Examples of performance data that should be captured during usability evaluation, and later analyzed to improve the product include:

- Timings (i.e., how long did it take to complete a task).
- Error rates for each task.
- The length of time it took to recover from an error.
- Steps omitted.
- Comparison of percentage of tasks completed correctly and tasks not completed correctly.
- Training or ramp-up time required to achieve competence.

Usability evaluations may range from paper-and-pencil evaluations and the collection of questionnaires and surveys, to actual usability tests.

TABLE 12.3
Sample Questions to Examine Usability

Usability Area	Sample Questions
General functionality	Did the user recognize and understand the range of activities available?
	Could the user visualize system use?
Information flow and screen navigation	Was the flow from one function or screen to another logical and smooth?
	Could the user select the appropriate module or functional area to complete tasks?
	Did the user try an inappropriate path?
Screen layout and object selection	Is placement of objects and menus consistent across screens?
	Can users develop a "mental map" of the system after working with a small set of screens?
	As the user works through a screen, are his or her movements fluid and logical (does layout support the work process)?
	Are objects and icons recognizable, reasonably sized, and easy to select?
Menu and command use	Are terms used to depict commands and actions relevant and understandable?
	Are the groupings of menu commands logical?
	Can the user identify any "missing" commands or functions?
Display issues and entry of data	Do screens support an appropriate entry order, enabling the user to move quickly through a screen?
	Are prompts or examples needed to guide input?
	Is text on the screen easy to read?
	Are items and information on the screen grouped logically?
	Is space used effectively to group information/tasks?
	Can the user discern what needs to be done to complete a task?
	Is information presented in a way that matches how the user views the problem and its solution?
	Are design elements (font, color, etc.) used consistently throughout the system?
	Can the user discern what to do, and where to enter information?

Paper-and-Pencil Evaluation

Using this technique, developers show users paper versions of screens or windows in a sequence and ask specific questions about them. Questions or activities typically are similar to these:

- Can you infer what each of these buttons might do by looking only at the icon?

- Are there any functions *not* represented by a button that you think might be necessary at this level?
- If I asked you to do <a task>, what do you think you would do first?
- Looking at this screen, what button would you select to do <a function>?
- Let's walk through the process of doing <a function or work process> and I will display paper versions of the screens that would be used. Then I will ask you questions about how well the screens support that process.
- What menu would you pull down to do <command>?

Data is collected for analysis as the evaluation proceeds. It is then used to determine how well user requirements have been captured, to determine additional requirements that should be added, and to refine the screens to enhance their usability.

Questionnaires and Surveys

Questionnaires and surveys are used to understand the preferences and beliefs of a broad base of users. The questionnaire or survey is considered a weaker form of information-gathering than the focus group. However, it is less costly, provides a means to access a larger group or sample, and enables the analyst to generalize to an entire user population. Surveys are generally used in the early stages to define the user requirements for a specific population. For example, they can be used to get the information that will create user profile models, including characteristics such as computer experience, job experience, and preferences.

Questionnaires may be used at any point in the development process, from task analysis and the verification of selected scenarios and demonstration scripts to usability evaluation. Any survey or questionnaire must be designed carefully and tested prior to use to ensure that the information obtained is of high quality and is useful in refining requirements. (In other words, "garbage in, garbage out.")

Usability Tests

Usability tests are conducted in controlled, simulated environments with small numbers of people (i.e., 10–12) who represent different portions of the end user population. The purpose is to address specific problem statements and capture data that will enable developers to work with a finite population identifying key opportunities to fine tune the software product *prior* to conducting acceptance tests or field studies.

Evaluation facilitators identify task components to be tested by reviewing objectives and critical success factors for the product, and by prioritizing what are perceived to be the most critical aspects of the software. Tests are presented to each user individually, their responses monitored, and target data collected. Later, evaluators determine the findings and make suggestions for refinements to the software product.

A similar approach is paired-user testing, in which two people use the prototype to perform a set of tasks together on a single computer (O'Malley, Draper, & Riley, 1984). The users are encouraged to discuss the prototype's behavior as they explore and interact with it, in response to specific written or verbal questions. This approach enables the facilitator to observe more and help less, which reduces the likelihood of bias.

Field Studies

A field study is a review of a prototype or demonstration system that has been placed in its natural setting. Often a project champion will allow the development team to install the system in his or her area. During the field study the project champion or other liaison captures data that can be used to further define requirements, especially in the area of usability. The types of data that may be collected during a field study include:

- Normal patterns of use.
- Functions used or accessed.
- Specific difficulties encountered (e.g., couldn't get to a function easily, couldn't enter data as easily as expected, etc.).
- Types of errors.
- Number of errors.
- Feedback on screen designs, including colors, icons, etc.
- Feedback on content.
- Missing functionality or content.
- Time to proficiency.
- Potential performance issues.

SUMMARY

Evaluation plays an important part throughout SEP—in each session conducted, each review of analyzed information, and in each review of storyboards, screens, and systems. Consequently, SEP enables developers to *build in quality* rather than simply "test" it in, because iterative evaluation helps ensure quality from both the customer's perspective and the developer's perspective. The fact that SEP is a user-driven, iterative, prototyping-intensive process (chapter 1) enhances the chances that requirements and system products will better meet the customer's needs.

Verification and validation occur in SEP at each step of the process. Users, shareholders, and project champions are frequently and continually involved with developers. They not only provide scenarios and information, but also review session reports, envisioned scenarios, screen designs, and prototypes or demonstration systems. Because products are developed based on information analyzed from session reports, it is possible to trace a requirement to specific session reports and sources. Traceability is helpful in ranking requirements, resolving conflicts, and ensuring that the software product reflects all identified requirements.

There are two primary SEP approaches to iterative requirements evaluation and refinement. The *proof-of-concept system* or prototype approach is used for projects that are reasonably sized and moderately well understood. The *demonstration system* is appropriate for very complex projects or very large domains. Regardless of the approach used, however, a variety of evaluation mechanisms are available. These include scenario reviews, reviews of session reports, storyboards/screens, demonstration scripts, focus groups, operational walkthroughs, paper-and-pencil evaluation, expert evaluation, field study, and usability reviews. Personnel involved in evaluation activities frequently include developers, analysts, users, shareholders, and project champions.

References

IEEE (1993). IEEE Standard Glossary of Software Engineering Terminology. Software Engineering Technical Committee of the IEEE Computer Society, IEEE.

Anderson, J. R. (1980). Cognitive skills. In J. R. Anderson (Ed.), *Cognitive psychology and its implications* (pp. 317–330). San Francisco: W. H. Freeman.

Anderson, J. R., Farrell, R., & Sauers, R. (1984). Learning to program in LISP. *Cognitive Science, 8*, 87–129.

Asur, S., & Hufnagel, S. (1993). *Taxonomy of rapid-prototyping methods and tools.* Paper presented at the IEEE 4th International Workshop on Rapid System Prototyping (June 28–30), Research Triangle, NC.

Auger, B. (1972). *How to run better business meetings.* St. Paul, MN: 3M Company.

Berlinger, D. C., Angell, D., & Shearer, J. (1964, August). *Behaviors, measures, and instruments for performance evaluation in simulated environments.* Paper presented at the Quantification of Human Performance, Albuquerque, NM.

Best, J. B. (1986). *Cognitive psychology.* St. Paul, MN: West Publishing Co.

Blomberg, J., Giacomi, J., Mosher, A., & Swenton-Wall, P. (1993). Ethnographic field methods and their relation to design. In D. Schuler & A. Namioka (Eds.), *Participatory design: Principles and practices* (pp. 123–156). Hillsdale, NJ: Lawrence Erlbaum Associates.

Boehm, B. (1981). *Software engineering economics.* Englewood Cliffs, NJ: Prentice-Hall.

Boehm, B., Gray, T., Seewaldt, T. (1984). Prototyping versus specifying: A multi-project experiment, *IEEE transactions on software engineering, vol. SE-10, 3* (May), pp. 290–303.

Bolton, R. (1979). *People skills.* Englewood Cliffs, NJ: Prentice-Hall.

Carey, J. M. (1990). Prototyping: Alternative systems development methodology. *Information and Software Technology, 32*(2), 119–126.

Carney, T. (1972). *Content analysis: A technique for systematic inference from communications.* Winnepeg, Canada: University of Manitoba Press.

Chi, M. T. H., Glaser, R., & Rees, E. (1982). Expertise in problem solving. In R. Sternberg (Ed.), *Advances in the psychology of human intelligence* (Vol. 1, pp. 7–75). Hillsdale, NJ: Lawrence Erlbaum Associates.

Christensen, J. (1950). The sampling technique for use in activity analysis. *Personnel Psychology, 3*, 361–368.

Cjelli, D. (1994). The FEFM Virus. *Performance & Instruction, 33*(10), 12–13.

Cooke, N. (1992). Modeling human expertise in expert systems. In R. Hoffman (Ed.), *The psychology of expertise* (pp. 29–60). New York: Springer-Verlag.

Coombs, C. (1964). *A theory of data.* New York: Wiley.

Cragan, J., & Wright, D. (1980). *Communication in small group discussions: A case study approach.* New York: West Publishing Co.

Dawes, R. (1964). Social selection based on multidimensional criteria. *Journal of Abnormal and Social Psychology, 68,* 104–109.

Drury, C. G. (1990). Methods for direct observation of performance. In J. R. Wilson & N. I. Corlett (Eds.), *Evaluation of human work* (pp. 35–37). London: Taylor & Francis.

Eberts, R., & Simon, N. (1984). Cognitive requirements and expert systems. In P. Salvendy (Ed.), *Human–computer interaction* (pp. 174–190). Amsterdam: Elsevier.

Enos, J., & Tilburg, R. (1979). Software design. In R. Jensen & C. Tonies (Eds.), *Software engineering* (pp. 87–89). Englewood Cliffs, NJ: Prentice-Hall.

Fishburn, P. (1974). Lexicographic order, utilities, and decision rules: A survey. *Management Science, 20,* 1442–1471.

Fleishman, E. A., & Quaintance, M. K. (1984). *Taxonomies of human performance—The description of human tasks.* New York: Academic Press.

Fox, W., & Glaser, R. (1990). *Improved nominal group technique trainer guide.* King of Prussia, PA: Organization Design and Development, Inc.

Frank, A. (1982). *Communicating on the job.* Glenview, IL: Scott Foresman.

Galloway, T. (1987). *TAXI: A taxonomic assistant.* Paper presented at the AAAI '87, Los Altos, CA.

Gammack, J., & Young, R. (1985). Psychological techniques for eliciting expert knowledge. In M. Bramer (Ed.), *Research and development in expert systems* (pp. 105–112). London: Cambridge University Press.

Glaser, R., & Chi, M. T. H. (1988). Overview. In M. T. H. Chi, R. Glaser, & M. J. Farr (Eds.), *The nature of expertise* (pp. xv–xxviii). Hillsdale, NJ: Lawrence Erlbaum Associates.

Gordon, S. (1992). Implications of cognitive theory for knowledge acquisition. In R. Hoffman (Ed.), *The psychology of expertise* (pp. 99–120). New York: Springer-Verlag.

Hackman, J. R. (1968). Tasks and task performance in research on stress. In J. E. McGrath (Ed.), *Social and psychological factors on stress.* New York: Holt, Rinehart & Winston.

Haddock, G., Kelly, J., Burnell, L., & Harbison, K. (1994). *Supporting concurrent engineering tools for the creation and reuse of component-based systems (Technical).* Arlington: University of Texas at Arlington.

Haddock, G., & Harbison, K. (1994, April). From scenarios to domain models: Processes and representations. In *Proceedings of Knowledge-based Artificial Intelligence Systems in Aerospace & Industry.* New York: SPIE: International Society for Optical Engineering.

Harbison, K., Burnell, L., Kelly, J., & Silva, J. (1995, July). *The scenario-based engineering process (SEP): A user-centered approach for the development of health care systems.* Paper presented at the 8th World Congress on Medical Informatics, Vancouver, Canada.

Hart, A. (1986). *Knowledge acquisition for expert systems.* New York: McGraw-Hill.

Hayes-Roth, F., Waterman, D., & Lenat, D. (1983). *Building expert systems.* Reading, MA: Addison-Wesley.

Hoffman, R. (1986). *Procedures for efficiently extracting the knowledge of experts (F49620-85-C-0013).* Bolling AFB, Washington, DC: Air Force Office of Scientific Research.

Hoffman, R. (1987). The problem of extracting the knowledge of experts from the perspective of experimental psychology. *AI Magazine, 8*(2), pp. 53–64.

Hogarth, R. (1974). Process tracing in clinical judgment. *Behavioral Science, 19,* 298–313.

Holtzblatt, K., & Jones, S. (1993). Contextual inquiry: A participatory technique for system design. In D. Schuler & A. Namioka (Eds.), *Participatory design: Principles and practices* (pp. 177–210). Hillsdale, NJ: Lawrence Erlbaum Associates.

Hsia, P., & Yuang, A. (1988, Jan.). *Screen-based scenario generator: A tool for scenario-based prototyping.* Paper presented at the 21st IEEE International Conference on System Sciences (pp. 455–461). Honolulu, Hawaii.

Hufnagel, S., Harbison, K., Doller, H., Silva, J., & Mettala, E. (1994, February 14–17). *National healthcare system information architecture: Using DSSA scenario-based engineering process.* Paper presented at the 1994 Annual HIMMSS Conference, Phoenix, AZ.

Hunter, S. (1985). *Texas Instruments AI Satellite Symposium.* Dallas: Texas Instruments.

Huseman, R. (1973). The role of the nominal group in small group communication. In R. C. Huseman, D. Logue, & D. Freshley (Eds.), *Readings in interpersonal and organizational communications, 2nd ed.* Boston: Hollbrook.

Jenkins, M. (1995, Oct.). The user's role in software quality. In *Proceedings, 5th International Conference on Software Quality* (pp. 366–374). Austin, TX.

Johnson, J. (1995). Chaos: The dollar drain of IT project failures. *Application Development Trends* (January), 41–47.

Jones, C. (1995, Oct.). Global software quality in 1995. In *Proceedings, 5th International Conference on Software Quality* (pp. 283–290), Austin, TX: ASQC.

Kahn, R., & Cannell, C. (1982). *The dynamics of interviewing.* New York: Wiley.

Kahneman, D., & Tversky, A. (1982). On the study of statistical intuitions. In D. Kahneman, P. Slovic, & A. Tversky (Eds.), *Judgment under uncertainty: Heuristics and biases* (pp. 430–454). New York: Cambridge University Press.

Kanal, L. N., & Lemmer, J. F. (1986). *Uncertainty in artificial intelligence.* Amsterdam: North Holland.

Kelly, J., & Harbison, K. (1995). Scenario-based engineering process: An object-oriented paradigm for complex system design. In *The scenario-based engineering process: Collected papers, (February).* Arlington: The University of Texas at Arlington.

Kelly, G. (1955). *The psychology of personal constructs.* New York: Norton.

Kirwan, B., & Ainsworth, L. (1992). *A guide to task analysis.* London: Taylor & Francis.

Klein, G. A. (1987). Applications of analogical reasoning. *Metaphor and Symbolic Activity, 2*(3), 201–218.

Klir, G., & Folger, T. (1988). *Fuzzy sets, uncertainty, and information.* Englewood Cliffs, NJ: Prentice-Hall.

Kolodner, J. (1983). Towards an understanding of the role of experience in the evolution from novice to expert. *International Journal of Man–Machine Studies, 19,* 497–518.

Kolodner, J., & Kolodner, R. (1987). Using experience in clinical problem solving: Introduction and framework. *IEEE Transactions on Systems, Man, and Cybernetics, 17,* 420–431.

Kurke, M. I. (1961). Operational sequence diagrams in systems design. *Human Factors, 3,* 66–73.

LaFrance, M. (1989). The quality of expertise: Implications of expert–novice differences for knowledge acquisition. In K. McGraw & C. Westphal (Eds.), *SIGART knowledge acquisition special issue* (Vol. 108, pp. 6–14). New York: ACM.

Lancaster, J. (1989). *Description of doctoral dissertation research in automotive bays.* Interview with Karen McGraw. Rosslyn, VA

Lawrence, B., Johnson, B., & Embley, P. (1995). Building an enterprise-wide defect tracking system: Lessons learned from learning lessons. In *Proceedings, 5th International Conference on Software Quality* (pp. 340–364). October 23–26, Austin, TX.

Lee, W. (1971). *Decision theory and human behavior.* New York: Wiley.

Likert, R. (1932). The method of constructing an attitude scale. In M. Fishbein (Ed.), *Readings in attitude theory and measurement* (pp. 90–95). New York: Wiley.

Lindwarm, R., & Norman, K. (1993). *AT&T teaching theatre (Research summary).* College Park, MD: University of Maryland Human–Computer Interaction Laboratory Center for Automation Research.

Mager, R. (1962). *Preparing instructional objectives.* Palo Alto, CA: Searon Publishers.

McGraw, K. (1991). *Videoconference consensus decision making session to refine quality engineer training requirements* [NASA Research Topic]. Houston, TX: Loral.

McGraw, K. (1992). *Designing and evaluating user interfaces for knowledge-based systems.* Chichester, UK: Ellis Horwood.

McGraw, K. (1994a). *Customer service workstation embedded performance support system cognitive task analysis (RWD-R-94155).* Columbia, MD: RWD Technologies, Inc.

McGraw, K. (1994b). Developing a user-centric EPSS. *Technical and Skills Training (October)*, 25–32.

McGraw, K. (1994c). Knowledge acquisition and interface design. *IEEE Software, 11*(6), 90–92.

McGraw, K. (1994d, Summer). *Performer-centric interface design.* Paper presented at the SHARE 82, Boston, MA.

McGraw, K. (1994e). Performer-centric interface design. *Performance & Instruction, 34*, 4, pp. 21–29.

McGraw, K., & Harbison-Briggs, K. (1989). *Knowledge acquisition: Principles and guidelines.* Englewood Cliffs, NJ: Prentice-Hall.

McGraw, K., & Riner, A. (1987). Task analysis: Structuring the knowledge acquisition process. *Texas Instruments Technical Journal, 4*(6), 16–21.

McGraw, K., & Seale, M. (1987, May). *Structured knowledge acquisition techniques for combat aviation.* Paper presented at the NAECON '87, Dayton, OH.

McGraw, K., & Westphal, C. (1990). *Readings in knowledge acquisition.* Chichester, UK: Ellis Horwood.

Metella, E., Harbison, K., & Hufdnagel, S. (1994, November). Scenario-based engineering process for reconnaissance, surveillance, and target acquisition. In *Proceedings of the ARPA Image Understanding Workshop* (pp. 1–16). New York: Morgan Kaufmann.

Michalko, M. (1991). *Thinkertoys.* Berkeley, CA: Ten Speed Press.

Miller, R. B. (1967). Task taxonomy: Science or technology? In W. T. Singleton, R. S. Easterly, & D. C. Whitfield (Eds.), *The human operator in complex systems.* London: Taylor & Francis.

Newell, A., & Simon, H. (1972). *Human problem solving.* Englewood Cliffs, NJ: Prentice-Hall.

Nierenberg, G., & Calero, H. (1971). *How to read a person like a book.* New York: Pocket Books.

Nisbett, R., & Wilson, T. (1977). Telling more than we can know: Verbal reports on mental processes. *Psychological Review, 84*, 231–259.

Norton, R. (1983). *Communicator style: Theory, applications, and measures.* Beverly Hills, CA: Sage.

Nunamaker, J., Applegate, L., & Konsynski, B. (1987). COLAB. *Journal of Management Information Systems, 3*(4), 27–38.

O'Malley, C., Draper, S., & Riley, M. (1984). Constructive interaction: A method for studying human–computer interaction. In *Proceedings of IFIP INTERACT '84: Human–Computer Interactions*, pp. 269–274.

Osborn, A. (1953). *Applied imagination: Principles and procedures of creative thinking.* New York: Scribner's.

Prerau, D. (1987). Knowledge acquisition in the development of a large expert system. *AI Magazine, 8*(2), 43–52.

Pressman, R. (1992). *Software engineering: A practitioner's approach.* New York: McGraw Hill.

Reboh, R. (1983, August). *Extracting useful advice from conflicting expertise.* Paper presented at the 8th International Joint Conference on Artificial Intelligence, Karlsruhe, Germany.

Reifer, D. (1985). State of the Art in Software Quality Management Report, Reifer Consultants.

Riner, R. (1982). *The ranking of job incumbents using CODAP overlap values to compare task inventories developed by a modified Delphi technique and a more traditional method.* Unpublished dissertation, Florida State University, Tallahassee, FL.

Ross, R. (1983). *Speech communication: Fundamentals & practice.* (6th ed.). Englewood Cliffs, NJ: Prentice-Hall.

Rubin, J. (1994). *Handbook of usability testing.* New York: Wiley.

Schermerhorn, J. R., Hunt, J. G., & Osborn, R. N. (1991). *Managing organizational behavior.* (4th ed.). New York: Wiley.

Schuler, D., & Namioka, A. (Eds.). (1993). *Participatory design: Principles and practices* (1st ed.). Hillsdale, NJ: Lawrence Erlbaum Associates.

Senn, J. (1989). *Analysis & design of information systems* (2nd ed.). New York: McGraw Hill.

Shadbolt, N., & Burton, A. M. (1989, April). The empirical study of knowledge elicitation techniques. *SIGART Special Issue on Knowledge Acquisition, 108,* 15–18.

Shannon, R. (1980, October). *The validity of task analytic information to human performance research in unusual environments.* Paper presented at the 24th annual meeting of the Human Factors Society, Los Angeles, CA.

Shaw, M. (1981). *Recent advances in personal construct technology.* New York: Academic Press.

Shiffrin, R. M., & Schneider, W. (1977). Controlled and automatic human information processing: Perceptual learning, automatic attending, and a general theory. *Psychological Review, 84,* 127–190.

Simon, H. A. (1981). Cognitive science: The newest science of the artificial. In D. A. Norman (Ed.), *Perspectives on cognitive science* (pp. 13–25). Norwood, NJ: Ablex.

Steep, R. E. (1987). *Concepts in conceptual clustering.* Paper presented at the 10th International Joint Conference on Artificial Intelligence, Los Angeles, CA.

Stefik, M., Foster, G., Bobrow, D., Kahn, K., Lanning, S., & Suchman, L. (1987). Beyond the chalkboard: Computer support for collaboration and problem solving in meetings. *Communications of the ACM, 30*(1), 32–47.

Steiner, I. (1972). *Group process and productivity.* New York: Academic Press.

Stewart, C., & Cash, W. (1985). *Interviewing: Principles and practices* (4th ed.). Dubuque, IA: Wm Brown Publishers.

Suchman, L., & Trigg, R. (1990). Understanding practice: Video as a medium for reflection and design. In J. Greenbaum & M. Kyng (Eds.), *Design at work: Approaches to collaborative design* (pp. 65–89). Hillsdale, NJ: Lawrence Erlbaum Associates.

Svenson, O. (1979). Process descriptions of decision making. *Organizational Behavior and Human Performance, 23,* 86–112.

Teichner, W. H., & Olson, D. E. (1971). Predicting human performance in space environments (NASA Contractor Report No. CR1370). *Human Factors, 13,* 295–344.

Thurstone, L., & Chave, E. (1929). *The measurement of attitude: A psychological method and some experiments with a scale for measuring attitude toward church.* Chicago: University of Chicago Press.

Tulving, E. (1972). Episodic and semantic memory. In E. Tulving & W. Donaldson (Eds.), *Organization of memory.* New York: Academic Press.

Tulving, E. (1983). *Elements of episodic memory.* Oxford: Clarendon Press/Oxford University Press.

Tversky, A. (1972). Elimination by aspects: A theory of choice. *Psychological Review, 79,* 281–299.

Van de Ven, A. H., & Delbecq, A. (1971). Nominal versus interacting group process for committee decision-making effectiveness. *Academy of Management Journal, 14,* 203–212.

Waldron, V. (1985). Process tracing as a means of collecting knowledge for expert systems. *Texas Instruments Engineering Journal, 2*(6), 90–93.

Waldron, V. (1986). Interviewing for knowledge. *IEEE Transactions on Professional Communications, PC-29 (2),* 31–35.

Wallace, A. (1972). Driving to work. In M. E. Shrio (Ed.), *Context and Meaning in Cultural Anthropology* (pp. 310–325). New York: Macmillan.

Wang, W., Hufnagel, S., Hsia, P., & Yang, S. (1992, June). *Scenario driven requirements analysis method.* Paper presented at the Second International Conference on Systems Integration, Morristown, NJ.

Whiting, B., & Whiting, J. (1990). Methods for observing and recording behavior. In R. Naroll & J. Cohen (Eds.), *Handbook of method in cultural anthropology* (pp. 282–315). New York: Columbia University Press.

Wielinga, B., & Breuker, J. (Eds.). (1985). *Interpretation of verbal data for knowledge acquisition.* New York: Elsevier.

Woodson, W. (1981). *Human factors design handbook.* New York: McGraw-Hill.

Zunin, L., & Zunin, N. (1975). *Contact: The first four minutes.* Los Angeles: Nash Publishing.

Author Index

Subject Index

Printed and bound by CPI Group (UK) Ltd, Croydon, CR0 4YY

17/10/2024

01775683-0008